Routledge Revivals

Office Development

Originally published in 1985, this book examines the impact financial institutions have on the location of investment of vast resources, including office development. An analysis of their behaviour is crucial to an understanding of the 20th Century urban development process. This book documents some of the international activity of property investment. Some cities in the UK, USA and France are examined in detail to demonstrate the huge physical impact of this development process. The constraints on office development are also discussed. A recurring theme is the power of the supply side of the development industry in comparison with the relatively weak position of the office end-user.

Office Development

A Geographical Analysis

Michael Bateman

Routledge
Taylor & Francis Group

First published in 1985
by Croom Helm Ltd

This edition first published in 2021 by Routledge
2 Park Square, Milton Park, Abingdon, Oxon, OX14 4RN
and by Routledge
605 Third Avenue, New York, NY 10058

Routledge is an imprint of the Taylor & Francis Group, an informa business

Publisher's Note
The publisher has gone to great lengths to ensure the quality of this reprint but points out that some imperfections in the original copies may be apparent.

Disclaimer
The publisher has made every effort to trace copyright holders and welcomes correspondence from those they have been unable to contact.

ISBN 13: 978-1-032-00538-6 (hbk)
ISBN 13: 978-1-003-17462-2 (ebk)
ISBN 13: 978-1-032-00547-8 (pbk)

DOI: 10.4324/9781003174622

OFFICE DEVELOPMENT

Office Development

A Geographical Analysis

Michael Bateman

CROOM HELM
London & Sydney

Croom Helm Ltd, Provident House, Burrell Row,
Beckenham, Kent BR3 1AT
Croom Helm Australia Pty Ltd, First Floor,
139 King Street, Sydney, NSW 2001, Australia

British Library Cataloguing in Publication Data

Bateman, M.
Office development: a geographical analysis.
1. Offices–Location
I. Title
338.6'042 HF5547.25

ISBN 0-7099-0697-8

Printed and bound in Great Britain
by Billing & Sons Limited, Worcester.

CONTENTS

TABLES

FIGURES

To Neil and Jonathan

PREFACE

Whilst the geographical study of office location has become an accepted area of study, there has been little detailed attention paid to the processes by which office development is created. The reality of most major cities, however, is that development of all property including offices, can be explained only by an analysis of the flow and location of the finance capital, which lubricates the development process. Most people are today aware of the power of the financial institutions involved in this process and indeed most have a vested financial interest in their activities. Yet few questioned would realise fully the spatial extent of their investment in the future via their pension fund contributions or their life assurance policies. A large part of today's urban development is based on this capital and it was a desire to demonstrate the spatial impact of this financial power which was the motivation for this book.

In the course of writing the book, I have inevitably questioned some of the accepted wisdoms of the property development industry. Its motivation is very different from my own, since its obligations are clearly towards those receiving pensions or those qualifying for them in the future, those anticipating the maturing of an insurance policy or the industry's shareholders. It requires perhaps, a detached commentator to analyse the impact and implications of its activities, especially as they relate to the transformation of urban areas. My obligations are less clear-cut but stem from a concern for the social life of cities and the framework within which it exists and flourishes.

An analysis of the effects of property development is no easy task given the difficulty of acquiring precise information on development projects. For this reason I am indebted to all of those both within the property profession and without, whose contributions have assisted me in constructing a composite picture of the development process. Whilst the opinions expressed and the errors are mine alone, I am extremely grateful to a number of individuals and organisations who have assisted me by giving information when requested. Particular mention must be made of Honor Chapman and her colleagues at Jones Lang Wootton in London and Paris, Michel Gleiser (IAURIF, Paris), the staff of Public Works, Canada, Cheryl M. Currie, (CIDC, Ottawa-Carleton), Paul Miller (A. E. LePage, Ottawa), Simon Fisher, Barry Quirk and Ron Hodrien. Many other organisations and individuals have been of indirect or direct assistance to me and I trust that they will accept a collective acknowledgement of my indebtedness to them. Sharon Jakobek painstakingly converted my manuscript into a legible typescript. Yvonne Court has not only shared my interest in this subject, but has patiently produced information gleaned from a wide range of sources and I am much indebted to her for her willing assistance. I must thank Peter Sowden for his encouragement to write this book and his patience in awaiting its completion. Finally I owe a debt of gratitude to my wife, Jen, for her unfailing support.

Michael Bateman

Emsworth

Chapter One

OFFICE DEVELOPMENT - SOME INITIAL CONSIDERATIONS

During the twentieth century, cities within western society have been characterised successively by the initial growth of office buildings, by their increasing functional and physical importance and by their dominant physical presence which has resulted in a transformation of many parts of the city, notably its central area. Many major cities, particularly those of North America are now recognisable by their skylines, composed largely of multi-storey office buildings. In Europe, skylines of public buildings have often been superseded by tower blocks of offices, rising up to overshadow towers and turrets of earlier centuries. Only occasionally have public authorities had the foresight to protect cities from a dramatic re-shaping and even then, measures have often been implemented rather belatedly. A process has been operating on cities throughout the developed capitalist world, encompassing Western Europe, North America, Hong Kong and Singapore, together with cities in the southern hemisphere including those of Australasia. The office resulting from this process not only exert a physical dominance, theyare also testimony to the increasing importance of office employment as a proportion of total urban employment. It is by no means unusual to find that white collar, or office, jobs account for over fifty percent of total employment in many major cities. Yet, despite the physical and functional importance of the office, with only a few notable exceptions, geographers have paid little attention to the phenomenon of office growth and development.

New construction in city centres almost always involves at least an element of office development and a city's growth is invariably accompanied by the expansion of its office sector. At the same time, this development activity provides a channel for investment funds from a wide variety of sources, including major institutional investors such as insurance companies and pension funds. Indeed most people in our society might claim an indirect ownership of modern air-conditioned office floor space, although its location may come as a surprise to

DOI: 10.4324/9781003174622-1

many, since investment is geographically extremely widespread. Investment in property is not made evenly across the urban system in either a spatial or a temporal sense. It is highly sporadic, favouring certain locations, at the expense of others, whilst it is also subject to a complex system of cycles of investment activity, linked to more general economic cycles. The resultant pattern of new office development is very uneven, with some cities suffering from a surfeit of new development, leaving others in which the stimulus of new office development would be welcomed by local politicians and planners alike, but which do not find favour with either investors or office users. A consideration of the emergent spatial distribution of office activity and of the factors which determine it, is long overdue and it is this which forms the central purpose of this book.

Certain fundamental questions need to be asked and attempts made to answer them. Amongst these, we may ask what mechanism determines that office development is actually set in motion. Leading on from this, what degree of control is exerted on the development process and how effective are the measures of control? How has office development affected urban areas, particularly those city centres which have been the focus of the most intense development activity? Within national urban systems, what has been the character of the geographical variation in the patterns of development? This question is part of a broader one which asks how it is decided that funds for office development are channeled into one particular urban system, given that the choice available may include not only cities within one country, but also numerous overseas locations. Finally, it is pertinent to ask how far new communications and information technologies are likely to affect the office development process. Once an attempt has been made to answer each of these questions, it becomes possible to suggest certain policy guidelines within which office development might reasonably be expected to operate.

GEOGRAPHICAL ANALYSIS OF THE OFFICE DEVELOPMENT PROCESS

In a general sense, studies of the office as an urban function came rather late, at least compared to those of other major urban employment functions, such as manufacturing, or retailing. Cowan (1967) offered an early analysis of the office as a major component of urban growth, but it was not until 1975 that Daniels produced the first major published analysis of the location of offices (Daniels, 1975). Goddard and Alexander have each analysed office location within a regional policy framework (Goddard 1975, Alexander 1979). Each of these reflected the growing awareness of the importance of the office in both an urban and a regional context, with Alexander adding a valuable international comparative dimension to the analysis. Nonetheless, despite these discussions of office location and development, from the point of view of a geographical anslysis, the intricacies behind the process of office development lay largely unexplored. One factor which goes some way to explaining this is that office development must be seen as the product of a development industry tied closely to the operation of western capitalist financial systems, parts of which are somewhat labrythine. Ambrose and Colenutt certainly made one very valuable attempt in their analysis of the so-called "Property Machine", in which some of the operations and requirements of those involved in office development are considered, although this discussion was confined to the intricacies of the British property industry and its domestic activities (Ambrose and Colenutt, 1975).

An understanding of current office development requires the development of a broader 'geography of finance', to include the operation of international money markets and international migration of investment funds. Dicken and Lloyd have already been amongst those who have emphasised the increasingly international dimension and interdependence of economic activity, and Taylor and Thrift's geographical analysis of multi-national companies further underlines the necessity to adopt an international view. (Dicken and Lloyd 1981, Taylor and Thrift, 1982). Certainly international variations in rates of exchange, interest rates for borrowers and interest returns for investors are now vital explanatory components in an understanding of international variations in economic activity.

The scale of operation of western capitalist organisations is such that only a consideration of their activities on a global scale and in the context of the world's financial system will yield satisfactory explanations for the results of their capital investments. Dicken and Lloyd go on to emphasise the role of major organisations, both public and private in shaping western society and environment (Dicken and Lloyd 1981). They point to the lack of geographical study of the finance sector, made all the more wanting since "a relatively small number of very powerful financial institutions act as *gatekeepers* in the flow of financial capital in industrial society." (Dicken and Lloyd 1981, p.65). Certainly there can be no satisfactory examination of the geography of property development, and particularly of office development without a detailed appraisal of, and familiarity with the workings of the financial institutions involved. It would be simplistic indeed to offer an explanation for office development in classical bid-rent theory terms, since utility maximisation by an occupier may be a secondary consideration for development to profit maximisation for the financial institutions. Although some would argue that the two may be equated, the notion of speculative property development by definition suggests a system where building activity is determined primarily by the supplier of property rather than by the eventual user. The major financial institutions play a crucial role in this process.

More generally the ownership of capital and variations in its availability are important to our explanation of geographical patterns of economic activity as is clearly demonstrated by Massey and Catalano's analysis of capital control and land ownership in Great Britain. (Massey and Catalano, 1978). Without such an analysis, an explanation of land use distribution and land use change is rendered at best imperfect and at worst, virtually meaningless. Yet an understanding of the urban development process is equally dependent on an explanation of the processes of property development, to include the flow of investment capital within the urban system and increasingly between the urban systems of different states.

It is clear that in the Middle Ages, cities developed in response to a given need, perhaps for trade, possibly for administration or some other major urban function, and in the Industrial Revolution, new buildings as well as entire towns were created to house new industrial processes. It is by no means clear today, however, that such a clear cut relationship exists, at least not in the short term. Speculative building for commercial functions, particularly for offices and retailing is now almost the norm, as well as being increasingly common for other functions such as industry and warehousing. It could be argued that historically, speculative urban development was confined to residential buildings and even then its history was relatively short, dating in the UK back to industrial housing of the nineteenth century. In the area of office building, however, there are a growing number of demands which bring about the process of development which are not directly linked to the needs and wants of the final user. Instead it has become a vehicle for large-scale investment of finance, some of which has been willing to accept long rather than short term returns. Indeed the property development industry is highly dependent on and compliant to the requirements of major financial investors. On occasions, the needs of user and developer will be immediately coincident in both time and space. Often, however, this is not the case and offices may be vacant for a considerable period of time. It is also quite conceivable that office activities, which are after all quite mobile, may move towards areas of available office space, in which case it is fair to claim as suggested above that it is the office development process as seen from the supply side which has shaped the urban system, rather than the demand from office users. It is a basic thesis of this book that the office development industry - part of Ambrose and Colenutt's "Property Machine" - which has been at least instrumental in transforming cities and in some cases has been the major guiding force.

DEVELOPMENT CYCLES AND OFFICE DEVELOPMENT

Development cycles in the urban system have been recognised for many years. Conzen's 'burgage cycles' are an early example of the recognition that urban form is the result of a continuous, but not necessarily smooth process of development. (Conzen 1960). The analysis of building cycles by Lewis, similarly recognised their importance in explaining the development process in a more general sense (Lewis 1965). Yet recent attention has focused on the more precise analysis of the relationship between temporal and spatial processes. Parkes and Thrift (1980), for instance, in acknowledging the importance of cycles of activity examine the alternative cycles which may be cited to explain variations in economic activity. The long term cycles of Kondratieff, covering very broad innovation periods have been briefly reconsidered by Hall (Kondratieff 1970, Hall 1981). It is difficult to discern the effects of such cycles in anything other than the broadest terms within the urban system. Yet shorter term economic cycles may indeed be readily identifiable in terms both of their general impact (see for instance Sant 1973) and of their more specific impact on urban form. (See Whitehand 1983, pp.51-53).

In the context of office activity, economic cycles have very obvious effects. The demand for offices is determined by the state of the business cycle. If it is in an upward stage then offices are in demand and are occupied, whereas on its decline, the reverse is true. By its very nature, however, the office development process is susceptible to such cycles of economic activity since it is incapable of short term adjustments to changes both in supply and demand. In terms of supply, office development projects may well take several years to come to fruition, during which time, rates of return on capital invested may change since rents may decline. Similarly if an office is planned and financed during the upward trend of the business cycle, but not finished until after the peak has been reached, then on completion there may be no tenant wishing to occupy the building. In either event, the expectation of rent income on which the building was financed will be unfulfilled. Should this happen, it is not uncommon for the developer of the building to keep a building vacant awaiting an upturn in the general economic cycle and hence a higher rent than may be obtained by letting it on a declining market.

Barras in his analysis of post-war development cycles in London suggests that the cycles themselves may be the product of the system of office development (Barras 1979 and 1984). Since there is a three or four year delay, he argues, between start and completion of a major scheme, development lags behind any changes in market conditions. An increase in demand causes rents to rise and a general shortage, leading to enhanced values and encouragement to developers to launch more schemes.

Once the first wave of schemes is complete however, shortages diminish, land values and rents level out, but some schemes are still left in the pipeline. Speculative development then ceases until the offices stemming from the last phase of the cycle are occupied. Certainly, it is not difficult to see the effect of such cycles in terms of oversupply following early stimulus. Paris in the mid-1970s was a spectacular example with over one million square metres of empty space, resulting from an over-stimulation of the local market at the end of the 1960s and in the early 1970s, coupled with a more general fall in demand, brought about by the post-1973 oil crisis induced recession.

In the case of London, the economic growth of post-war Britain sustained demand until the mid-1960s, when there was an oversupply. As Barras points out, since that time, the business cycles and development cycles have been out of phase since short-lived economic or business cycles have meant that offices generated on the upturn have been completed only after the peak of the cycle. This situation was especially well-illustrated in the mid-1970s when sudden economic collapse caused a major oversupply of space, It also set off a decline in property values so severe that it caused a major re-shaping of the property development industry as companies' book of assets of property fell dramatically, resulting in some spectacular bankruptcies of property companies which had financed new development on the basis of loans secured by property which had fallen dramatically in value.

The international dimension of the property development industry assumes considerable importance in the context of these cycles of development. Whilst it is true that there are world-wide economic cycles, fuelling or depressing demand for office space in a general sense, there are also spatial variations which can be seen in terms of the operation of these cycles. The case of Houston is a good one with its development cycle boosted because of its links with the energy industry in 1973 and after. A general re-appraisal of the desirability of dependance on distant resources and a domestic development of the energy industry resulted in considerable development for Houston as an office centre. The City saw frantic activity in its office sector during the 1970s so that by the beginning of the 1980s, Houston had much enlarged office stock. In 1981, 1.05 million square metres of new offices were let, a level 250% above its closest rival, Los Angeles. (Estate Times Review, January 1982).

In contrast, Paris and Brussels, two cities which will be discussed further in later chapters, both saw increasing vacancy rates during the same period, since demand for offices declined in the face of economic recession. By the 1980s, Houston was suffering from an oversupply of offices and was entering the downturn of its development cycle. This brief example illustrates that it is unreal to talk of only one development cycle. Instead, each national urban system has its own cycle, determined by such factors as national economic performance, and rates of international currency exchange. Moreover, even within one national urban system, there may exist simultaneously several development cycles, each affecting different cities. The USA provides an example of this as will be seen when the North American office market is further discussed in Chapter Seven.

Such temporal variations in development activity assume a special significance in the context of the office development process since much of the capital involved is highly mobile. In the case of British capital, the lifting of exchange controls in 1979, following the election of a Conservative government, led to a large scale movement of capital overseas. At around the same time, British investors withdrew much of their assets from European cities and invested in property in the USA, the effects of which are discussed in later sections of this book. Temporal swings in exchange rates also affect the channeling of property investment funds. The decline in the value of the Belgian franc in the late 1970s caused disinvestment from that property market since the real value of rents to foreign-based property investors fell. Similarly, British and other investment in French property has been at a very low level because of the low value of the franc, even within the European Monetary System, in the last two years, i.e. 1982-84.

The final example of temporal constraints on investment relates to the relative value of alternative investments. Property as an investment has always had the virtue of security, but rates of return on property are not as high now as they were in the late 1970s. Indeed as Table 1.1 shows, returns to two types of institutional property investors have recently fallen quite sharply.

Table 1.1: Annual Rates of Return on Property Investment, 1976-83

% yr. to Dec.31	1976	1977	1978	1979	1980	1981	1982	1983
Phillips & Drew Property Unit Trust Index	5.0	28.1	19.2	26.2	16.2	14.9	7.2	8.2
Phillips & Drew Pooled Property Fund Index	7.4	25.8	19.9	23.1	18.7	16.3	8.3	7.6

Source: Phillips and Drew, quoted in Investors Chronicle, 6 April 1984.

Once alternative investments offer a much higher return than that offered by property, then the level of institutional finance channeled into property will obviously fall. The recent past has seen such a situation, although it should be emphasised that the position is somewhat volatile and the long-term security of property may still be attractive. Nonetheless, recent surveys have suggested an annual return from property investment of 8.2% in 1983, compared to 28.9% for equities and 10% for cash investments. (Financial Times, 24 January 1984). As a result, institutional investment in new property fell by 30% during the first nine months of 1983 compared to a similar period in the previous year. Thus temporal changes in the relative attraction of investment opportunities affect the availability of finance for new urban development, including that destined for offices. Development capital is not therefore, a fixed supply, but is in practice highly variable over time. It must be said, however, that despite recent trends, the cost of disinvestment from property is very high, especially within a declining market and as the following chapter points out, property holdings by major financial institutions are high and likely to remain so given the past record of property as a secure long-term investment.

SWITCHES AND TRIGGERS

Nonetheless this ability on the part of investors to decide to place money in any one of a number of investment opportunities illustrates well the notion of "switches" as discussed by Dicken and Lloyd (1981, p.102). They suggest that the development "process is channeled through two 'switches'". One switch is seen as determining whether funds go into property or some other investment outlet, such as the Stock Exchange, whilst the second, once money has been channeled to property, determines whether the investment goes into British or overseas property. We have considered above some of the factors which determine the operation of these switches, but we may further complicate the analogy by suggesting that there are also important 'trigger' mechanisms, which can alter the flow of investment funds, to divert them either to an alternative property market or to an alternative investment outlet.

Many such triggers will emerge and be discussed in later chapters, but we may note by way of example certain activities which may have an important and discernible 'trigger' effect. Certainly for overseas investors in property, the devaluation of a currency may mean that rental returns are so diminished as to make continued investment if not entirely unprofitable, at least not as profitable as alternative investments. Conversely, revaluation may have an opposite effect. Legislative measures may also act as a trigger to development activity either by initiating it or by causing its sudden cessation. The imposition of rent control on the commercial sector would be a case in point. Even the threat of such controls has been sufficient in the past to deter investment in property. The lifting of foreign exchange controls in Britain in 1979, already mentioned above, was an effective trigger to increased development activity overseas by British interests. Political events, such as a change of political persuasion of a government may be very important. The socialist victory in France in 1981 certainly did little to stimulate the property development industry, at least as far as foreign investment was concerned, although it has to be admitted other factors also played their part. Political events such as the British prime minister's visit to China to discuss Hong Kong's future may be sufficient to shake confidence in a local property market, as will be seen in Chapter Four. Triggers, as either explosive stimuli or sharp brakes on development must therefore be added to switches to aid our understanding of the complexities of the office development process. The switches are operated according to decisions taken within the context of cycles of economic activity, but a trigger event may interrupt the process quite rapidly.

CURRENT APPROACH TO AN ANALYSIS OF OFFICE DEVELOPMENT

In broad terms, the approach adopted in this analysis is to set the context of office development by examining its instigators and the controls within which they work, before examining specific office markets. The structure of the property industry is discussed in Chapter Two. The relationship between property companies and the financial institutions has changed greatly since the early property 'boom' of the post-war period in Britain and this evolving relationship is considered in this chapter. The current role of financial institutions in promoting office development is extremely important and indeed the size of their property holdings is very considerable. Legal and General, for instance, has 31.4% of its £7,160 million of investments in property, whilst the largest British portfolio is that of another insurance company, the Prudential, which owns property valued at approximately £3,500 million. It is not only British finance which has funded the property development industry, since for instance, approximately 20% of French institutional funds are invested in property, whilst in the Federal Republic of Germany, there is a legal obligation on financial institutions to invest 20-30% of their investment funds in property. Nonetheless, this analysis will focus on the British property industry - a parochialism which may be excused by the power and influence of British property interests which will become evident in other sections of this book.

Controls on both the development process and the investment process are considered in Chapter Three, since neither market is totally unfettered. Instead they operate within a framework of controls which range from the restriction of foreign land ownership, common to many countries at the present time, to local planning directives. Developers have to operate within local and regional planning policy guidelines and these are considered in this chapter together with restrictions on the flow of investment capital. All such controls are important in explaining the manner in which the office development market is directed.

Some cities have been quite markedly transformed as a result of the operation of the office development system. In Chapter Four, two such cities, Brussels and Hong Kong, are considered in detail. In both cases control mechanisms existed to direct development. In the case of Brussels, it was *via* normal planning legislation, but the operation of planning was extremely lax, leaving an almost anarchic pattern of new office development activity to transform the city. In Hong Kong, the control has been in theory more direct since the government determines the flow of land on to the market for property development. Yet in practice, office development and the considerations of the market made a very considerable impact on the city. The result in both cities was a rapid expansion of office development with only secondary consideration being given to its impact on the social relationships and physical form of the city. Particular office markets are the subject of Chapters Five, Six and Seven. Whilst in the case of Britain, the London market is seen as being dominant, the British provincial market is nonetheless far better developed than that of France where the office market of the Ile-de-France (Paris) region far outweighs that of the rest of the country. At the outset of 1984, for instance, there was a stock of new or renovated office space of 1.9 million square metres on the Paris market, and 133,600 square metres in Lyons, whilst no other French city had office stocks in excess of 100,000 square metres (Bourdais, 1984). Whilst Chapters Five and Six examine the British and French markets respectively, Chapter Seven concentrates firstly on the USA and then on Canada. The USA has a very varied market and has seen a considerable influx of overseas capital to finance property, with the result that some cities, such as Denver and Houston (already briefly discussed above) have seen dramatic rises in their office stocks in very short periods of time. Canada is selected for some scrutiny since its capital, Ottawa, offers a detailed case study of the direct effect of central government policies on a local office market.

The last twenty years has seen considerable advances in firstly communications technology and then more recently in information technology. Both areas have potentially profound implications for office development. Enhanced telecommunications may in theory at least, offer considerable locational choice to office users. Introduction of integrated systems of information storage and retrieval on the other hand may fairly be described to be revolutionising the operation of many offices. At the same time, the physical requirements of these new technologies raise important questions concerning the suitability of older office buildings and hence their economic viability. Chapter Eight endeavours to explore some of these questions, whilst admitting that in this particular area, a degree of speculation concerning the future is almost inevitable.

Finally attention is turned to some of the public policy implications arising from the issues discussed in this analysis. The role of public authorities in controlling and directing office development is very varied if one takes an international view. Similarly, as has been evident in Britain, controls have varied over time. Yet it is difficult to escape the conclusion that the interests of the entire community are so intricately tied up with those of the office development industry that a clearly defined public policy is required.

Data Sources - an explanatory note

Monitoring of the operation of the office development system is somewhat difficult. The gathering of original field data whilst often possible may often not yield information concerning the financing and the ownership of office properties. Important secondary sources of data are therefore indispensable in analysing the office market. In this case, the property pages of major newspapers, and especially of the Financial Times, have proved to be extremely valuable sources of information. Additionally the major journals of the professions involved in property development contain much valuable information concerning individual property developments. Thus journals such as the Estates Gazette and the Investors' Chronicle have yielded data which is specifically acknowledged in the text. The continuing custom of those involved in property development in Britain to deal in imperial rather than metric measurements is unfortunate, especially since data is often presented in rounded form, e.g. an office development of 50,000 square feet or an office stock of 6.5 million square feet. Where such rounding has evidently taken place in the original data source, both the metric equivalent and the original imperial measurement has been included in this text.

Chapter Two

THE PROPERTY MARKET

Urban areas have been undergoing constant change, but none more dramatic than the transformation of their centres with large amounts of new office buildings during the past three decades. The extent of this change in fact varies quite widely, reflecting not so much variations in the demand for office space *per se*, but the aggregate decisions of the suppliers of new office space. There is little use, therefore, in seeking explanation for these changes in classical accounts of urban land use distribution and change such as bid rent theory, if only because most such theory was developed prior to the introduction of new conditions and processes affecting urban property and land use in the post-1945 period. It is important to realise at this point that change in the urban area in the form of new buildings does not take place spontaneously in response to a market demand, derived from the abstract notions of urban land economics. Instead, it must be seen as the result of a complex series of deliberate investment decisions on the part of a range of people, many of whom have little concern with the classical notions of location and accessibility, but are far more dictated to by requirements of secure returns on financial investments. As such, market experience, intuition and a willingness to take measured risks may be more important to the decision-maker than a knowledge of urban land use theory. As we shall see, market demand may itself be a highly notional concept with change in urban centres being carried out in anticipation of a perceived market demand.

A second important underlying precept must also be understood if the rapid growth in office floorspace is to be totally explained. It is simply that in the post-war period, and especially in Britain, there has developed a financial structure which is vitally dependent on the inherent value of land and property. Offices then are merely recipients of investment finance directed in order to ensure maximum security and profitability. The unspoken assumption that neither property nor the land on which it is built, is likely to lose in value, is a powerful motivating factor in this process,

DOI: 10.4324/9781003174622-2

although it was also an assumption which of necessity was closely re-examined in Britain during the mid-1970s. The operations of the financial markets in Britain are such that new property development, and especially offices, may often be seen as merely by-products of the financial structure, provided by a process which at its inception may be operating almost independently of the demand for the new floorspace.

We cannot proceed to an analysis of office development without taking with us a clear picture of the processes at work in promoting and encouraging property and particularly office development. Often the operation of such processes has determined both the form and timing of office development to an extent which far outweights the influence of either planner or ultimate user of the building. Whilst the following analysis of the property market is largely centred on the operation of the British property market and the behaviour of British financial institutions, there are sound reasons for this apparent insularity. The British property market has been extremely active, in the post-war period, far more so than that of any other European country. The processes do have an international dimension, however, since those involved in British property development have extended their sphere of activity into other countries. Thus the rapid growth of offices in Brussels in the early 1970s, the recent vicissitudes of the Hong Kong property market and the explosive office development of Houston since 1970 have all involved finance, developers, investors and letting agents originating in the UK. Indeed, in many cases, the British role goes far beyond mere involvement since the property development cycle itself has been initiated and maintained by the activities of British interests. The focus therefore of all of this activity in terms of finance, management expertise, and decision-making is London. Hence it is not a narrow parochialism which determines the shape and content of this chapter, since the repercussions of the British property market have been felt much further afield.

PROPERTY AS AN INVESTMENT

Property, and especially offices, has long been regarded as a secure form of investment. Not only does it provide a security in terms of maintaining the real value of assets, but it also may provide an income through rent revenue, and possibly an increasing asset through periodic rent increases. It also has, in a general sense, a degree of liquidity which may be attractive to an investor. The limited supply of land has led to a time-honoured view that unlike other limited, but exhaustible assets and commodities, it offers complete security since it cannot be lost or over-exploited over time. Whilst it is true that other investments may well offer higher rates of return on capital invested, they may carry with them a higher degree of risk than investment in land or property. In the post-war period, rapid inflation also dictated that property should become a focus for investment funds. Its record, whilst not entirely without blemish, was very good in staying ahead of inflation.

Whilst its growth in value since 1976 has not matched inflation, if taken in the long term, it is still a comparatively secure investment. Table 2.1 indicates the rate of return on property investment compared to investment in equities as measured by investment by specialised funds. Plainly the recent performance does not compare well with equities, which have been on a rising market, but its ten year performance is superior to that of equities.

Table 2.1: Medium Rate of Return of Pooled Funds (% per annum)

	One year to end - 1982	One year	Ten years
Property Funds	9.1	14.3	13.5
Equity Funds	29.4	15.2	12.0

Source: The Wyatt Company Survey of Pooled Pension Funds, quoted in Investors Chronicle, 25 March 1983.

The secure return was of special importance to a finance sector which was expanding rapidly on the basis of new forms of 'with profits' insurance and index-linked pensions. Even the occasional lapse in the growth of its value, such as occurred in 1974, was soon forgotten and faith in its worth quickly reconfirmed. Thus, one important decision-maker in the property market, M. H. Mallinson, the Joint Chief Surveyor of Britain's largest property investor, Prudential Assurance, was able to re-affirm his faith in property in 1983. "I cannot do other than believe that the reality of real estate still makes it a desirable and valuable commodity in a world of unreality". (Mallinson, 1983).

The property market is dominated by financial institutions with a very large amount of money to invest annually - £2.18 billion in 1981 and £2.6 billion in 1982 (Financial Times 29 April 1983). The financial institutions include insurance companies and pension funds as well as unit trusts specialising in property. Interest in land in the form of direct ownership by these financial institutions is by no means new. D. Massey and A. Catalano have examined the development of this phenomenon, demonstrating, however, that until the twentieth century, their direct ownership of land and property was small. (Massey and Catalano 1978, p.123). Until the post-1945 period, however, their interest was usually *via* mortgage finance rather than direct ownership of land. Such interest retains the security of investment for the institutions, but has the potential of only limited returns through interest repayments, whilst the capital growth of the value of the land and property accrues to the mortgagee. Such a situation was not in the interests of the financial institutions and from a constantly evolving relationship between financial institutions and property developers in the post-war period, a very different pattern of interests in property has emerged in which the interests of the financial institutions has become central to the property market.

The changing relationship in the post-war period can be explained by a number of important factors. Firstly, the growth of the long term life assurance business needed to be matched by long term investment opportunities. Investment in land and property offered one such avenue for investment which had the added attraction of offering a relatively high level of security. This was not the only avenue to be explored, but it did grow to the extent that by 1982, insurance companies invested £1.05 billion in real estate, approximately 21% of their total investments in that year. Massey and Catalano cite the Pilcher Report to underline the fact that liquidity was of less importance than security of income and indeed demonstrate that property investment became such an accepted part of the insurance companies' overall asset portfolio that, "because of the steady annual growth of premium income, it has rarely, if ever, been necessary for institutions to realise property assets in order to meet liabilities... they have come, in effect to regard their investment as permanent." (Pilcher Report quoted by Massey and Catalano (1978), p.124).

More recently such permanent holding of property assets has not been a characteristic of the property folios of financial institutions such as the insurance companies. Instead there has been a considerable growth in the trading of properties involving intra-institutional deals as well as deals with property developers. All such dealings are aimed at maximising both returns and long term security, but in themselves emphasise a further attraction of property - its marketability within the well-established property market system.

The growing and increasingly active participation in the post-war property market owes something to a second factor stemming from the situation in early post-war Britain. W. Lean and B. Goodall in an early analysis of the British property market point to the part played by the leaseback transaction, (Lean and Goodall, 1966). This was essentially a device by which companies could use property to raise capital. The latter was the subject of tight controls exercised by the Capital Issues Committee so that post-war reconstruction could be controlled, to some degree. A loophole in the regulations, however, permitted the leaseback transaction by which a company could sell its interest in a property to a third party, but continue to use it by leasing it back from the third party. For the growing funds of the insurance companies, this presented an excellent form of secure investment. It is interesting to note that the device is by no means out of use and some retailing chains in the UK have entered into similar agreements with financial institutions involving valuable 'high street' properties in the recent past in order to raise capital. Post-war inflation meant that insurance companies were increasingly turning to investments other than fixed interest shares. During the 1950s, the general pattern was for insurance companies to finance property indirectly through property companies, but by 1960, they were seeking a greater degree of equity participation in any company which they helped to finance. The process continued so that "by May 1962, the insurance companies held ordinary shares in nearly fifty property companies, and by 1965, the figure has risen to over a hundred." (Lean and Goodall, 1966, p.94).

The property companies themselves grew rapidly in number, from only twenty-five quoted companies in 1939 to over 150 by 1965. Their growth was rapid and it was to continue unchecked until the abrupt reversal of the market which began in late 1973. In the mid-1960s, however, property companies dominated the market, often financed by the so-called institutional investors, i.e. the insurance companies and pension funds. The latter were attracted to property investment for many of the same reasons which had attracted the insurance companies. Particularly for funds requiring steady and secure growth to ensure that pensions could always be paid in the future, preferably on an index-linked basis, short term liquidity and potentially high, but speculative profit were quite secondary to considerations of long term security and steady growth in the value of any investment.

Property as a secure asset has been and continues to be particularly favoured. Its underlying security has been dramatically underlined during the economic recession of the early 1980s by property transactions in the retail sector in the UK. In conditions in which consumer spending has been in relative decline, many retailers have been forced to look carefully at their property assets with a view to gaining a degree of liquidity. Thus, as noted earlier, companies have entered into leaseback transactions to raise capital, based solely on the value of their retailing property. The grocery and supermarket chain, Tesco have entered into a number of such transactions, whilst J. Hepworth and Son, the multiple tailors, began in 1982, a programme of property disposals to reduce its borrowings. (Investors Chronicle, 25 March 1982, p.24). A further example is the leaseback purchase of D. H. Evans department store by Legal and General in 1981 for £29 million. (Plender 1982, p.36). The attraction of such transactions can be seen from table 2.2 which gives an indication of the level of returns on property values. The returns are actually quite varied, but illustrate the fact that some retail companies may be better placed with their property assets realised and more working capital to earn greater returns.

Table 2.2: Property Assets and Return on Values of Major British Retailing Organisations

	Property Assets (£m)	% Free-hold	% annual return on property values
British Home Stores	136.9	n/a	31.0
Currys	88.0	n/a	11.9
Debenhams	261.2	67.9	8.8
House of Fraser	330.5	63.6	11.4
Marks and Spencer	1,040.0	n/a	21.0
Sainsbury	287.9	53.4	28.9
UDS	256.3	61.7	5.3
F.W. Woolworth	483.9	65.4	9.7
Tesco	267.7	50.2	15.9

Source: Investors Chronicle, 25 March 1983

It is not surprising therefore, that recent bids for the major chains of UDS and F. W. Woolworth in the UK., have been largely inspired by the property assets rather than the trading units themselves. It seems likely that the resilience of property as an investment will continue to ensure its predominant position with investors of all scales. The investors themselves are the vital decision-makers on whom property development and therefore urban changes are dependent. It is therefore appropriate to continue our analysis of the property market with a close examination of each of the investors.

FINANCIAL INSTITUTIONS AND PROPERTY INVESTMENT

The property market is dominated by pension funds, insurance companies and property unit trusts, all of which are direct investors in property on a large scale. Property development companies are also active, although their role has changed recently, and their direct investment is now far less important in the property market than in the early post-war years when they acted as an indirect channel for investment of the funds of other parties such as the insurance companies. Insurance companies and pension funds act directly with funds acquired as premiums or pension contributions. Property unit trusts on the other hand, are more indirect agents of investment since they exist either to enable small pension funds to acquire an interest in a range of property thus offering the security of a well balanced and diversified property portfolio, or to enable larger funds to invest, particularly in overseas property, offering management expertise which might not otherwise be available. Average holdings in such trusts are less than £0.5 million, indicating the limitations under which any small pension fund would be operating should they be acting individually in the property market. We may add to these investors, the banks, since they are a source of funding for property. Their participation may be indirect since they may well fund both property companies engaged in speculative commercial property development of all types as well as residential development companies on a large scale. It is also evident from Table 2.3 that in some years, banks have been withdrawing from property investment. Table 2.3 summarises the net investment in property by UK companies from 1976-1982.

Table 2.3: Net Investment in Property by UK Institutions, 1976-1982 (£m)

	1976	1977	1978	1979	1980	1981	1982
Insurance Companies	449	410	549	628	881	1053	1059
Pension Funds	513	534	591	498	855	756	731
Capital Market*	19	-12	-2	80	147	97	258
Property Unit Trusts	71	66	109	90	77	108	55
Sub-Total	1053	998	1247	1296	1960	2014	2103
Banks	-58	-166	-326	-190	61	283	770
Total	955	832	921	1106	2021	2297	2873

*Funds raised by Property Companies on the stock market.

Source: Vickers da Costa, United Kingdom Research, Property Sector Review 23, August 1983.

The volume of investment shows a considerable recovery from the mid-1970s, when confidence in the property market was at an extraordinarily low level. The increasing participation by the banks indicates this renewal of confidence. However, the recent trends have been somewhat less encouraging for the property market with institutional investors seeking alternative investment, such as foreign government securities and equities. The pattern of investment by pension funds in 1982 vividly illustrates this as the four quarterly investment figures for that year were successively £228 million,

£210 million, £195 million and £98 million. The total investment of £731 million was in some contrast to the £897 million two years earlier. Yet, despite the short term trends, the level of investment is still impressive. In 1981, insurance companies invested 16% of their invested funds in property and in 1982, despite the falling trend, pension funds still invested 11% of funds. (Vickers da Costa, 1983). The recent downturn should, however, be put into the context of a remarkable rise in investment during the 1970s. Pension funds invested £97 million in 1970, (13.2% of funds), £339 million by 1975 (15.3%) and £897 million (13.6%) by 1980. (HMSO 1982). The growth of investment by insurance companies was similarly very noticeable, with £197 million (24.8% of funds) invested in 1970, rising to £855 million (15.8%) by the end of the decade. Any fall in the early 1980s was therefore from a very high level of investment compared to ten years previously.

A closer analysis of each of the investors will further explain some of the fluctuations shown in Table 2.3. In addition, the changing role of each major agent will be clarified. At this stage we may merely note that when these investment figures are translated into new urban buildings, we are considering very powerful agents of large scale urban change.

THE INVESTORS

a) Property Companies - a declining power

There can be little doubt that the property companies in the UK were at the forefront of the post-war urban development process. Others were to imitate their methods making adjustments to suit particular circumstances. It is equally true that the British property companies brought about a fundamental change in attitude towards the urban development process in neighbouring countries of Western Europe, setting the example of speculative building on a large scale. By the 1980s, property companies had acquired a distinctive role which was complementary to the development activities of the large financial institutions. This state of affairs was, however, no more than a final stage in a process of continuing adaptation to market conditions on the part of the property companies throughout the last thirty years.

In the 1950s and 1960s property companies developed buildings and held them as investments. In the early years, there was no shortage of sites, particularly in bomb-damaged London. The abolition of building licenses in 1954 provided a further stimulus to their activity. (Plender 1982, p.91). Their finance was from the insurance companies, content to see their money in the traditional security of property. Naturally the property companies had to pay for their finance, but interest rates were low and profits were to be made even when capital and interest repayments were subtracted from rental income. The higher interest rates of the next decade, the 1970s, brought about a change in practice. Property companies could not hold onto their developments and still pay off loan charges. In that event, the common practice which arose was for the property development companies to sell off their completed buildings, often to the financial institutions. The profits of the property company were then derived directly from the sale of the completed building and not from rental income. The change in practice meant in a general sense, that there was a change in ownership of new buildings - particularly office buildings. From the point of view of the financial institutions, it brought their interest in property to a more direct level although a transitional phase in which insurance companies put funds into secondary banks who in turn supported the property developers was important until the market crash of December 1973. There were other factors which caused the institutions to become more commonly the owners of buildings in their own right. Certainly wage inflation had pushed up pension funds' assets, but at the same time, the depressed state of the market meant that the institutions could acquire a fixed range of property. The assets of liquidated property companies were also on the market. At this time for instance, the ICI Pension Fund acquired for £19 million, the Victoria Centre, a major retailing development in Nottingham, whilst BP's Pension Fund paid £45 million for the Knightsbridge Estate. This period was a difficult one for property companies and in the period after 1974, those which survived the collapse of the market sought a new relationship with the institutions. They required stronger financial backing, especially for the higher risk projects with which they were being increasingly associated. This often entailed the formation of a partnership between a property company and a major financial institution such as a pension fund.

The natural following stage was for the financial institutions to want a larger share of the development profit. Inevitably it would have been logical to foresee a final stage which would see the elimination of the property company since it had become little more than an intermediary between the financial institutions and the full development profit. Yet this has not happened, largely because the property company still has a vital role to play, particularly in the higher risk, possibly peripheral areas, where their experience of the development process may be vital for a successful development. A high risk development would not endear itself to the more conservative financial institution, but a property company retains the potential of making high profits, provided it is willing to risk the failure of the scheme. In some ways, therefore, the property company has fallen from its pre-eminence of the 1960s. The rise and fall of property development companies is reflected in the number of such companies quoted in the London Stock Exchange. In 1958 there were 111, by 1964 the number had risen to 183 after which there was a slight decrease by 1967 to 164. By 1977 however, the effect of the market crash of 1974 had been felt and only 105 were still in existence, a downward trend which was to take their number down to around ninety by the 1980s. (Moor 1979). Those which remain however, still have an active part to play in the development process both in the UK and elsewhere.

b) The institutional investors

The two most powerful investors are undoubtedly the pension funds and insurance companies. These total assets available for all forms of investment of all kinds has risen from £7 billion in 1957 to £85 billion by 1981. (Plender 1982, p.14). The rise of the financial institutions as urban land owners has far eclipsed the importance of individual land ownership. In a more general sense, institutional ownership of assets has replaced ownership by individuals. It could be argued that since a large proportion of the population have insurance policies and pay into pension funds, that ownership is shared widely, albeit indirectly. But the essential fact remains that decision-making in terms of urban land development, particularly in the commercial sector, is increasingly concentrated in the institutions. J. Plender in his thorough review of the development and power of the institutions, points to the "depersonalisation of capitalism". (Plender 1982).

It is difficult to over-emphasise the size and hence the power of the financial institutions. A steady in-flow of funds is a characteristic of their *raison d'être* and results in a continuing need to find investment opportunities. Not all such opportunities are in property - Plender calculates that 80% of ordinary share capital in the UK is now owned by the institutions, compared to 28-36% in 1939. Even if only a small proportion were to be continually directed into urban development, however, it would be sufficient to ensure a continually evolving urban fabric simply because the insurance companies and pension funds do have huge resources at their disposal. An example is provided by the Post Office Pension Fund, the largest in the UK., which had in 1980, an inflow of capital of £1.5 million, each working day. The National Coal Board Pension Fund had assets "far in excess of the book value of all the assets of Britain's state-owned coal mining industry." (Plender 1982, p.15). Table 2.4 indicates the largest of the institutional investors, listing those with investments in excess or approaching £2 billion.

Table 2.4: Investment by Major British Insurance Companies and Pension Funds in 1983

	Total Investments (£m)	Property Holdings (£m)
Prudential	13,240	3,460
Legal and General	7,160	2,250
Commercial Union	5,860	666
Standard Life	5,680	971
Norwich Union	4,900	1,410
Post Office PF	4,750	1,310
Royal	4,710	571
Eagle Star	3,510	450
National Coal Board PF	3,370	1,170
Sun Alliance	3,120	754
Guardian Royal Exchange	3,100	611
Electricity Supply Ind. PF	2,800	700
General Accident	2,470	485
British Rail PF	1,927	391

Source: Estates Gazette, 1 October 1983.

To this list can be added the British Government's Crown Agents and the Church Commissioners. The investment managers of all these institutions wield very considerable power important to the urban development process. The desirability or otherwise of this situation is of course open to debate and powerful arguments have been advanced against the all-pervading influence of the institutions and the consequent centralisation of the decision-making which affects the urban environment. Plender consistently presents the argument against this situation, (Plender 1982), whilst R. Cowan writes "an increasing proportion of the nation's capital is becoming controlled by a smaller number of professional managers and trustees of a small number of financial institutions. Unlike companies which are controlled by their shareholders and regulated by statute, the pension funds are remarkably free to use their enormous resources exactly as they wish. That would be a matter for no concern if it were not that the investment decisions of the institutions have important consequences for the economy and the environment." (Cowan 1982, p.192).

British financial institutions are not the only ones to invest in the property market, since in other European countries, a similar pattern of investment prevails. Yet there are important differences between the UK and other European countries. In the Federal Republic of Germany, there is a legal obligation for 20-30% of funds of the institutions to be invested in property. (Hillier Parker, 1983). Approximately 20% of French institutional investment is in property. In both cases, however, the tradition has been to invest in the residential sector rather than the commercial sector, whereas British institutions' portfolios have usually been almost exclusively devoted to commercial and industrial property with the greatest emphasis on offices. The Swiss government imposes considerable restrictions on the operation of financial institutions, requiring a proportion to be invested in property, but permitting neither overseas investment by Swiss institutions, nor the acquisition of property in Switzerland by foreign institutions. The institutions of Dutch origin act in a manner quite similar to those of the UK., including a proportion of overseas investment such as in the US property market. The pension funds in the USA also invest in property although it was not until 1974 that their pattern of investment could be said to be similar to that of the British financial institutions (see Chapter 7).

c) Property Unit Trusts

Property Unit Trusts are the final major investors in property. They have become very active since their introduction in Britain in 1966, although their total investment is small when compared to the institutional investors, as table 2.3 indicated. It is also true to say that the 1980s has seen fluctuating fortunes for the property unit trusts, resulting, for instance, in a fall in net investment in property from £108 million in 1981 to £55 million in 1982. Purchase of units in property unit trusts is limited to approved pension funds and charities which are exempt from income and capital gains tax under the provisions of the 1970 Finance Act. Their operation is closely controlled by the Department of Trade - they are not permitted, for instance, to advertise for new unit purchasers. Originally they were a useful vehicle for small and medium-sized funds by which they could enter the property market. During the 1970s, larger pension funds began to purchase units in the property unit trusts, especially those which were specialised in certain areas such as North American property. An alternative method for the smaller funds is to be party to pooled pension fund investments often managed by major life assurance companies.

The funds themselves vary greatly in size, with fewer than fifty unit-holders in the smaller trusts, whilst the largest, the Pension Fund Property Unit Trust (PFPUT), has over 600 holders. Average holdings in the larger trusts is in the region of £400,000. (James 1983). The advantages of the property unit trust for all pension funds is that they offer the opportunity to spread investment funds in property both geographically and by sector. Thus a small fund which, acting independently may be limited to the purchase of one provincial office block in Britain, can invest in property in a number of US cities to include both office and retail holdings. Naturally the unit-holders also benefit from specialist management and investment experience. The trusts also offer a liquidity which is lacking in the holding of property directly, since units can be withdrawn from the trust, although in times of a weak property market, redemption offers may be fairly low.

Property unit trusts have faced problems recently. For instance in 1982, the best gain of any property unit trust was 13.6%, whilst the largest and first established trust, the PFPUT gained only 1.1%. (Investors Chronicle, 25 March 1983). The weak performance of property in 1982 caused a high level of withdrawals from the trusts, resulting in the drastic reduction in net investment between 1981 and 1982 cited above. This led to some degree of rationalisation of this sector of investors, such as the linking of the PFPUT to the property management interests of Morgan Grenfell, the merchant bankers, to form Morgan Grenfell Property Services. (Financial Times, 20 May 1983). The PFPUT had been forced to sell 7-8% of its property portfolio in order to maintain an acceptable liquidity ratio in the face of withdrawals. By 1983, however, the market was improving with relatively little disinvestment from property unit trusts. Table 2.5 shows those unit trusts when funds in 1983 were valued at over £1 million.

Table 2.5: Property Unit Trusts with Holdings Valued in Excess of £1.0 million

	Fund valued at 1.3.83 (£m)
Fleming Property Unit Trust	274.0
Pension Fund Property Unit Trust	247.1
Hazard Property Unit Trust	179.5
Local Authorities Mutual Investment Trust	127.1
Hill Samuel Property Unit Trust	116.9
American Property Trust	112.0
North American Property Unit Trust	100.0

Source: Adapted from Investors Chronicle p.45, 25 March 1983

The table indicates that some property unit trusts are highly specialised in terms of their investments. This is true both geographically and by property sector. The most common geographical specialisation is in North America, which has generally enjoyed a high degree of confidence by British property investment agencies in the 1980s and especially since 1982. The activities of these unit trusts is explored further in Chapter 7. The other property sector for which specialised property unit trusts exist is the agricultural sector, although in general the funds in this area are not large.

RECENT TRENDS IN THE PROPERTY MARKET

The long term security of investment in land and property interests must not disguise the fact that there may well be cycles of relatively low profitability or poor performance when compared to other investments. This cautionary note is particularly important in assessing the recent past record of rents in the property sector. Naturally the growth in real terms in rents accruing from property is very much tied to the general health of the economy. In Britain, the period from 1982-83 was one of some uncertainty as to whether the economic recession had truly "bottomed out" and therefore, whether economic recovery was in its early stages. This uncertainty was reflected in rent levels and also in the comparative performance of property on a regional basis since on the one hand, prime office building in the south-east metropolitan area was let easily whilst industrial building in the north-east remained in the market with little prospect of any rent return at all, let alone rental growth.

Late 1982 saw a significant fall in property values and strong competition from other investment opportunities. As indicated earlier, however, in a discussion of Table 2.1, the short term has not been good for property but it still holds its own over the last ten years compared to alternative investment opportunities. The picture is of course complicated by the annual rise of inflation since if it is low, commensurately smaller rent increases can produce better returns on investment in real terms. The graphs in Figure 2.1 showing the Investors Chronicle Hillier Park Rent Index (a commonly used trade barometer of rental growth), demonstrate this clearly, since in Figure 2.1A, the trend is downwards, but when the rental changes have been adjusted for inflation, as shown in Figure 2.1B, an upward trend, albeit slight, emerges. Finally one should note that rents are not independent of supply and a fall in construction may well mean that rents will increase in the long term. (Investors Chronicle 25 March 1983, p.17).

The geographical implications of such temporal changes in returns are clear. Certain favoured locations, the prime sites within cities, and more generally central cities themselves are likely to become greater foci for property investment at the expense of more marginal locations. The core-periphery contrast may become very marked as investment managers in a professional role care little for the social and environmental considerations of investing in towns and cities which show little sign of sharing in economic recovery. A similarly conservative approach will be reflected at the international level, when the riskier investments in secondary centres, such as Southern Europe may well be rejected in

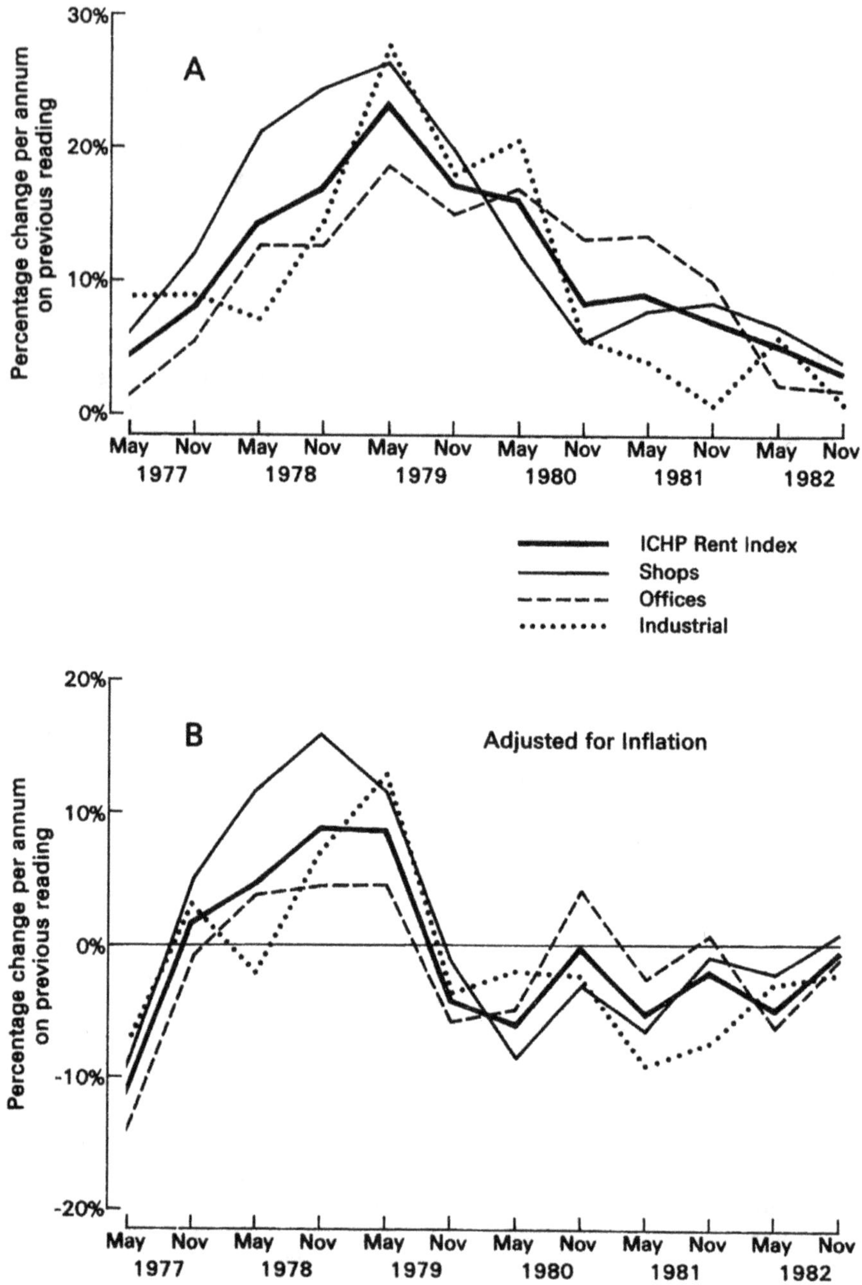

Fig. 2.1 Investors Chronicle Hillier Parker Rent Index

favour of the known returns, which may be modest but are assured, of investment in Frankfurt, Paris or Houston.

It is nonetheless true that this activity tends to be a part of a system of self-fulfilling prophecies on the part of the investing institutions themselves. Since the financial institutions are the major holders and buyers of property, it follows that trading in property tends to be an internalised operation. "Property is, to all intents and purposes, a one-way market in which the institutions create their own capital values." (Investors Chronicle 25 March 1983, p.9). The favouring of any particular site, therefore, has often as much to do with mutual confidence in it as a secure location for investment as any inherent characteristics of the location itself, although admittedly they cannot be completely discounted. If enough institutions have investment in one centre, however, that in itself ensures it a certain degree of security for further investment. Only occasionally does that particular axiom fail, although when it does so, the results can be spectacular. The recent history of the Hong Kong property market is an alarming example of this (see Chapter 4).

The uncertainty of the recent past has caused changes in the behaviour of institutional investors. Property investment by life assurance and investment funds in the first three quarters of 1982 fell to 13.5% of net cash flow compared to 14.5% in the same period in 1981. There were by early 1983, other signs of at least a rationalisation of property holdings by pension funds and large life assurance companies in response to predicted returns on property investment, especially when compared to alternative investment opportunities. On the other hand, there is no escaping that it is still the financial institutions who are the major owners and developers of new commercial land use in cities and it is decisions in their boardrooms which determine the shape and dynamics of urban areas, and especially the central areas of cities so dominated by offices. All that planners in any country can do is to guide each such change, since only rarely can they make a more positive contribution.

Reviewing the post-war period, it is possible to see a series of 'trigger' mechanisms which have unleashed investment in property and subsequent change in the urban fabric. The restriction on capital issue in the post-war period and the subsequent development of the leaseback transaction was one such device which ensured that insurance companies acquired an interest in property. More dramatically in terms of promoting development, the lifting of exchange controls in 1979 by the new Conservative government in Britain opened the doors for property investment overseas on a hitherto unknown scale.

INTERNATIONAL PROPERTY INVESTMENT

Those countries which have been the recipients of British investment on a large scale have frequently seen a transformation of their property markets. The most spectacular example of this is Belgium with the Brussels market in particular undergoing remarkable fluctuations as British investors first invested on a large scale in the early 1970s, only to withdraw later in the decade. A full analysis of the Brussels market is contained in Chapter 4.

Nonetheless, there is little continuity in the temporal pattern of investment in that few countries have consistently found favour with British investors in property. Figure 2.2 shows the pattern of investment in property by British companies in three European markets and the USA in the period 1972-81. The Belgium/Luxembourg market showed a small level of net investment flow until the late 1970s. The change to disinvestment was not surprising since by then the British induced property boom, particularly affecting Brussels, had produced a large-scale over-supply of offices. Other negative factors included the fall in value of the Belgian franc, depressing recent returns for British investors. Net earnings from the market were actually negative in 1977 (-£7.7 million), 1978 (-£3.1 million) and 1979 (-£0.5 million), with net profits only returning in 1980 (£2.4 million) and 1981 (£2.3 million). The disenchantment with the Belgium market (Luxembourg investment was always negligible compared to that flowing into Belgium), was by the late 1970s widely felt, resulting in selling of property and a sharp disinvestment, amounting in 1981 alone to a withdrawal of £25.1 million.

Although the detailed pattern is not precisely the same, a similar trend was noticeable in France, where the peaks of investment came slightly earlier and there was not the large scale disinvestment from the property market. The Netherlands, on the other hand, has never had such a strong attraction for British investment and following net losses of £16.5 million in 1977, the rate of disinvestment from that market was very noticeable with withdrawals of £32.0 million and £33.0 million in 1978 and 1979 respectively.

In sharp contrast to the downward trends in the investment flow to Western Europe during the 1970s, the North American market attracted a remarkable increase in British funds invested in property. The North American market is virtually the USA market, since Canadian Government restriction on overseas investments embodied in the Foreign Investment Review Act (1973) mean that the flow to that country is relatively small and net earnings low (a total net earning of £2.6 million from 1977 to 1981). The US market trend is actually understated in Figure 2.2 since the figure for 1980 is not available in the official statistics. The 1979 level of

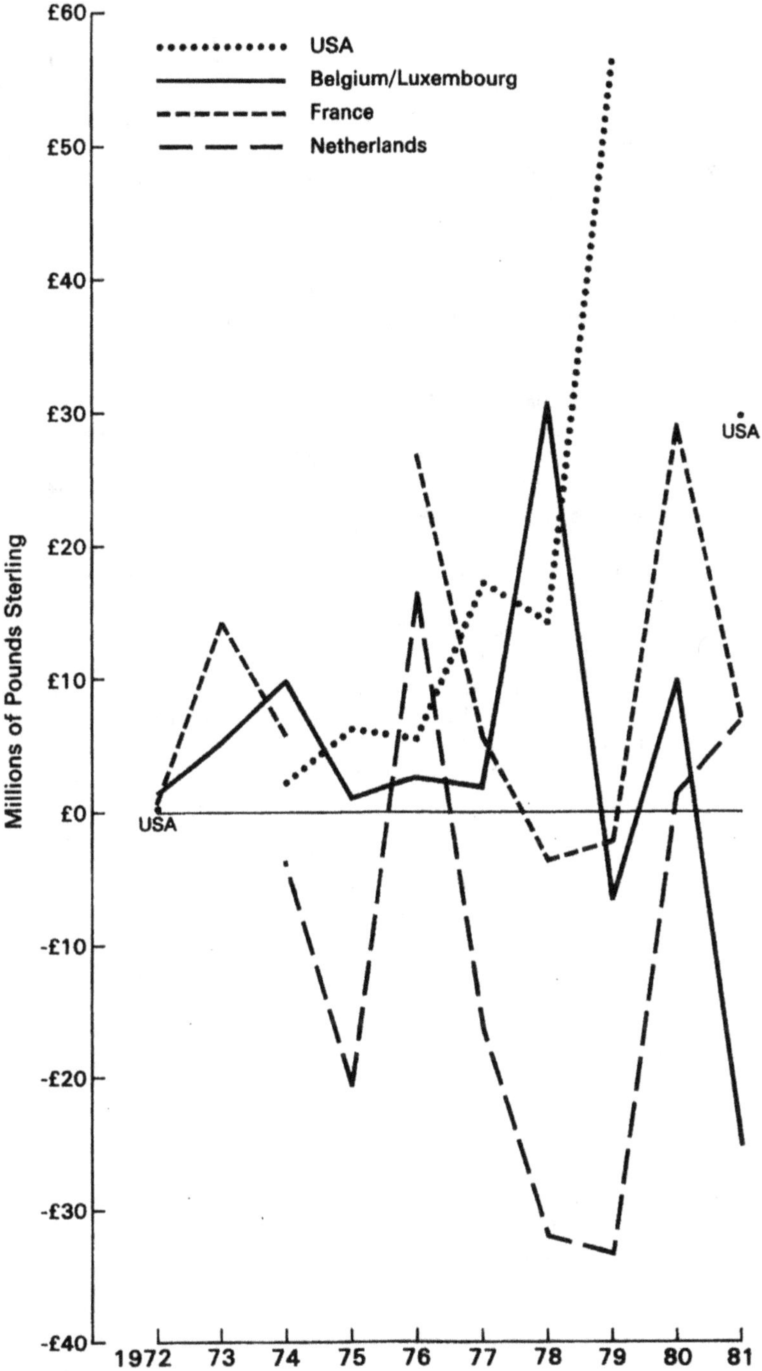

Fig. 2.2 Investment of British Funds in Selected Overseas Property Markets 1972 - 1981

£56.4 million was without doubt vastly overshadowed by a figure of at least £250 million in 1980 alone. Since the total for North America was £290 million and it can safely be assumed that Canadian bound investment (with a disinvestment of £6.2 million in 1979 and an investment of only £7.0 million in 1981) could have formed only a small proportion at the most of this huge sum. The lifting of exchange controls in Britain during 1979 no doubt lubricated the outflow of funds generally in 1980, but this in itself can hardly explain the sudden and dramatic rush to the US property market. It could be argued that investment managers were reluctant to be left out of a growing market, and joined a North American band-wagon. On the other hand, such investment decisions were based on good performance of North American property as an investment in the late 1970s. Average yearly net earnings to the UK were £7.78 million for the period 1977-80, compared to an average loss of £2.68 million on EEC property investments during the same period. (HMSO 1983). As the analysis of the US property market in Chapter 7 indicates, the massive flow of funds into the USA was not spread evenly across all cities. Instead, initial investment favoured the East Coast market, concentrating on New York, whilst by 1980, it was the expanding and relatively free markets of Houston and Dallas which were becoming the major foci for British investment funds. A more complete picture of the importance of British investment in overseas property is obtained if it is realised that the return investment in property by foreign concerns purchasing land and property in the UK is negligible by comparison to the outward flow of funds. In 1980 there was a net outward flow £376.6 million by British concerns investing in overseas property, whilst the net return flow of foreign investment was only £1.2 million. Comparable figures for 1981 were a net outward flow of £164.5 million and a net return of £23.3 million, a slightly more balanced pattern but still with an overwhelming emphasis on a net outflow of property investment funds.

An analysis of the property portfolios of individual property companies underlines the trend discussed above. Table 2.6 shows the proportion of property holdings of four major companies held in various overseas markets. The table is derived from a monitoring of property companies carried out by a British merchant banking organisation.

Not all property companies have invested large sums overseas. For instance, Capital and Counties, whose property assets have increased greatly since 1977, still have a portfolio of properties which is overwhelmingly based in the UK. On the other hand, some companies have a very large proportion of their assets overseas. One example is the Hammerson Property Investment and Development Corporation PLC whose increasing Australian interests between 1980 and 1983 took its investments from £95 million (19.0% of total investments) to

Table 2.6: Distribution of Property Holdings by British Property Companies 1977-83

			UK	Europe	N.America	Australia
Capital & Counties	1977	%	97.1	-	2.9	-
		£m	67	-	2	-
	1980	%	97.5	-	2.4	-
		£m	119	-	3	-
	1983	%	98.1	-	1.8	-
		£m	212	-	4	-
Hammerson	1977	%	41.8	5.5	23.8	29.0
		£m	137	18	78	95
	1980	%	55.0	0.4	26.0	19.0
		£m	275	2m	128	95
	1983	%	52.3	-	16.1	31.7
		£m	486	-	150	295
MEPC	1977	%	59.0	13.1	19.2	8.6
		£m	310	69	101	45
	1980	%	79.5	9.0	4.8	6.8
		£m	550	62	33	47
	1983	%	63.8	8.0	12.5	15.9
		£m	649	81	127	162
Slough Estates	1977	%	67.6	10	15.7	6.7
		£m	142	21	33	14
	1980	%	78.7	7.4	9.3	4.3
		£m	296	28	35	16
	1983	%	73.0	4.4	16.4	6.3
		£m	360	22	81	31

Source: Vickers da Costa Ltd., Property Quarterly Reviews.

£295 million (31.7%). This is partly explained by the acquisition of a minority shareholding in the company by an Australian financial institution between 1980 and 1983 since the Australian government, like that of Canada, has imposed restrictions on investment by non-Australian companies. MEPC and Slough Estates, are examples of companies which have a very broad geographical spread of property assets, of which the European proportion has been steadily declining in contrast to other markets, particularly Australia and North America.

The possession of overseas property by the financial institutions is undoubtedly very high. A recent survey showed that whilst UK property investment by the pension funds was falling substantially, the flow of funds overseas was increasing. (Financial Times, 13 January 1984). Of the funds surveyed with assets in excess of £160 million, 69% of overseas property assets were in the USA, 6% in West Germany, with 2% in the Netherlands and Belgium. The vast majority of these investments were offices.

The property investment market is complex in its operation, particularly since it has acquired an overseas dimension. The channeling of money to promote new urban development is determined not by need or demand, but by the relative profitability of alternative investments. The alternatives maybe in sectors other than property, such as industrial equities, or they may be other geographical locations. The precise destination of the property investment will be decided by a variety of factors including for instance, exchange rates and political decisions and events. Such factors add to the difficulties to be faced in analysing the operation of the property market. Nonetheless, an understanding of the market and an appreciation of its role is vital to a geographical appraisal of the office development process.

Chapter Three

CONTROL OF THE OFFICE DEVELOPMENT PROCESS

The interests of the property development industry are largely served by constructing new buildings within cities in order to secure a return on investment. The industry is not, however, unfettered in pursuit of this goal. In a general sense, developers of offices and their occupiers have to conform to the constraints of the policies and planning legislation which are operative in whatever areas they are considering the development of new office space. Whilst this may seem a rather obvious statement, it is important to realise that the nature of control in the office development process varies greatly, as does the effectiveness of the actual control measures. There can be little doubt that office development interests have been attracted to areas of relatively weak control, such as Brussels in the early 1970s and Houston since the early years of that decade. Understandably developers have been more reluctant to fight their cause in places in which control is more severe.

The development process in its international context is also constrained by governmental restrictions which are often made on overseas investments. These are considered in detail in the second part of this chapter.

THE NATURE OF CONTROL POLICIES

Control policies may attempt to place restrictions either on the supply of offices or on the demand for them by potential occupiers. In practice, both national and local governments have usually favoured the former with a system of permits and authorisations being required by an office developer prior to the construction of the building. On the other hand, there are powerful arguments in favour of the latter type of control system, involving either occupation permits or taxes particularly if the aim of the control is to move office functions away from congested cities. In such a case, the

DOI: 10.4324/9781003174622-3

office employer may be asked to pay what amounts to a congestion tax, which is then seen as a part of the operating costs of the office. If the employer decides that such a tax makes the operation of his office uneconomic then he would move his operation elsewhere. The long-term effect would be to deflect the office development itself into areas where the economic penalties were lower for the office user. Such control is, however, indirect in terms of its effect on the office development industry and there are few examples of the policy being implemented without supplementary controls of the development process itself. Nonetheless, the control of occupiers rather than developers does have the merit, of making office users examine very carefully their need for congested locations. Far more common, however are controls on the location and supply of offices. This brings us to a consideration of the role of planning in the office development process and the degree to which the market is directed and determined by public authorities.

The fact that it needs to be controlled is in itself revealing of the nature of office development since it suggests that what is beneficial for the property investment interests is not necessarily perceived as being of a more general benefit to the community at large. Nonetheless the degree to which governments have been willing to acknowledge this and attempt to exercise control and restraint has varied, both over time and space. Thus, whilst some local authorities have been willing to accept the benefits of large scale office development, no matter what the cost, others have been far more circumspect. We may point for examples, to the contrasts in policies of Croydon in the 1950s and certain inner London boroughs in the 1980s. Croydon's development was initiated by the Croydon Corporation Act 1956, allowing a measure of municipal ownership of land in the centre of the town, amounting to two acres in all, (Daniels 1975, p.201). Early development of this area and adjacent land already in public ownership, together with a later allocation of 45 acres in local authority ownership resulted by 1971 in the development of over 500,000 square metres of offices before the process began to decelerate. The headlong rush into office development by the local authority resulted in severe social and economic transformation of the areas actually built upon, coupled with a more general alteration of the role and nature of the town itself. Indeed by 1982, it had 1.53 million square metres of offices, the largest office centre outside central London and in excess of Manchester (1.50 million square metres) and Birmingham (1.34 million square metres). A permissive attitude, albeit within the framework of planning legislation, thus allowed the office development industry to have very much its own way. A contrast is provided by inner London boroughs in the 1980s, such as Lambeth and Southwark. As will emerge in later discussion (see Chapter 5), their policies towards office development have been infinitely more restrictive. The social penalties of office development in failing to provide employment for local residents, whilst often displacing small local industrial employers, have been recognised, resulting in policies which discourage the activities of office developers.

The relationship between the planning system and the property market is complex and often difficult to analyse fully. N. Moor writes that:

> inevitably the property developer finds himself pitted against his chief obstacle: the planning authority. The authority determines in principle where and what quantity of floorspace the developer can construct, but the relationship is more subtle than this since the developer generally has one major advantage in that he is the initiator, and can attempt to change policies with regard to location and quantity of floor space. (Moor, 1979, p.194).

The situation envisaged by Moor is very much that of the British office market, although similar conditions prevail elsewhere. The planning authority can provide the context for development and can prohibit development through the development control system. Yet it would be often politically unacceptable to refuse all development. Neither is it realistic to expect the planning system to create offices according to published plans since however else it may be interpreted, it is a servant of the office development market. Local planning authorities can only rarely act as developers in their own right. Furthermore, a persuasive developer may convince a planning committee of the benefits of a particular development despite the fact that in theory it transgresses policy guidelines set by the planners themselves.

Elsewhere, outside the UK, the planning system itself may be very much weaker, allowing the developers to determine for themselves the nature and direction of development with little interference from planning authorities. Brussels during the early 1970s offered these advantages to the development industry and it responded in a spectacular fashion, transforming sectors of the city in a short space of time before a degree of control was imposed (see Chapter four). Similarly the dramatic increase in office space in Houston has been attributed to the fact that there is virtually no control. Jones Lang Wootton in their assessment of Houston see the situation somewhat differently since there is a form of zoning with "land use controls which are based largely upon private contracts or so-called restrictive covenants, or in other words, upon master planning by the private developer/ investor". (Jones Lang Wootton 1980). The control is therefore somewhat different from that exercised in many other cities both in Europe and North America and the relative lack of public interference in the development process increased its office inventory by an annual average of 418,000 square metres (4.5 million square feet) during the 1970s, to produce a total by 1979 of 6.69 million square metres, 75% of which was added from 1975-79. Office developers were certainly attracted by such an environment, buoyed by the city's petroleum industry and one in which developments could come to fruition relatively rapidly after inception.

OFFICE CONTROL IN THE UK

One of the most sweeping control measures on office development was that which operated in the UK from 1964-79 involving Office Development Permits (ODP). Its overall effect has been well documented elsewhere (see for instance Alexander (1979) pp.63-70), but it is pertinent to reconsider it in the present context. The system was introduced in late 1964 in an attempt to curb development of offices with a particular aim of stopping the excesses of speculative development in central London. The product of a Labour Government, the system included an embargo on development in London and it became a more general regional policy measure for a short time when restrictions on office developments were by 1966 extended to the South East region, the Midlands and East Anglia, although the last region and rural parts of the Midlands were to be excluded in 1969 and by 1970 only the South-East remained under the control system. The phenomenal growth of offices in London in the 1950s and 1960s required such control measures since it had led to increasing problems associated with congestion and journey to work. Between 1948 and 1963 over 500,000 square metres of office space received planning permission in the central area of London and even by 1957, the London County Council had been forced to produce its plan to control congestion in the capital. By the late 1950s, growth had spread to other centres in London, including such suburban centres as Croydon underlining the necessity for controls in development.

The aims of the policy were therefore to relieve Central London of its congestion and to attempt to redistribute office development away from the capital firstly into other centres in the South East and then more generally beyond the region. (Alexander 1979, p.65). The establishment of the Location of Offices Bureau (LOB) in 1963 enabled companies to gain advice should they wish to relocate physically from the capital. The bureau acted purely in an advisory role, and companies were not forced to seek its advice. Nonetheless, it was instrumental in assisting the relocation of a considerable number of office jobs from London during the 1960s, its own estimate for the period 1964-69 being 84,000, although

R. Hall (1972) indicates a somewhat different figure, perhaps as high as 150,000. The LOB itself was disbanded in 1979 at the same time as the total lifting of ODP controls in Britain. In 1977 it was given the rather difficult task of attracting new office development to inner urban areas, including those of London, where there was also to be an accent on attracting overseas companies to the city. In addition, a regional role was spelt out for the bureau to promote a more even distribution of office employment in England and Wales. The election of a Conservative government in 1979, however, heralded its ultimate demise.

It could be argued that by then LOB had performed its task of assisting the deconcentration of offices from London. It was certainly not true however that growth of the office sector in the capital had ceased. In the period 1965-77, 1,399 ODPs were issued for Central London, permitting over 7 million square metres of new office floorspace, a net increase of 4.6 million square metres. Indeed, as Alexander points out, the annual increase in office floorspace, permitted in the London region rose from 0.3 million square metres in the period 1948 to 1963, to 0.6 million square metres from 1963-76. (Alexander (1979), pp.66-67). Nonetheless, the intra-regional redistribution was important with less than half of the latter period's permissions being given for the central area, compared to 60% in the 1950s.

The demand for new office jobs was therefore hardly constrained by the office development permit system. Certainly this was hardly surprising given that the control was not on the potential occupiers of new office space, but on the developers. A system of occupation permits, such as that operated in the Paris region in 1969 for occupation of premises of more than 100 square metres, would have been rather more effective in restricting office growth to only those companies which could demonstrate a need to be located in the capital. In the event, certain functions showed very considerable growth tendencies in London during the 1970s, including the foreign banking sector. This particular activity had grown in importance from 73 such banks in 1960 to 158 in 1970. From that date however, its growth was spectacular, to 356 by the end of the decade, and 428 by 1982. The growth rate of foreign banks in the city was in the 1970s, and early 1980s running at the level of one new establishment every two weeks. (Noel Alexander Associates 1983). Such growth, both here and elsewhere in the finance and business service sector proved difficult to restrain and it was not particularly surprising that a Conservative administration chose to lift the controls which had operated since 1964.

Nonetheless, it should not be thought that the policy of redistributing office development was a total failure. Certainly at the intra-regional level, some success was achieved as can be seen by a brief analysis of one planning area, in the

South East, Central Berkshire. The publication of the Strategic Plan for the South East in 1970 (HMSO 1970), outlined certain major growth zones in South Eastern England. By that time, the ODP system had been modified so that only projects in the South East required a permit. The Strategic Plan in outlining major growth areas nonetheless had to depend on a government policy of sustaining ODPs in order for its provisions to be achieved. The policy as originally outlined foresaw the need to disperse to areas in the South East, and thence to other areas in the UK. Local authorities in the South East, such as Berkshire, interpreted the policy to mean that areas in the South East, designated in the Strategic Plan were to be favoured. Thus, the Report of Survey of the Central Berkshire Plan states "the aim of ODP policy is to divert office employment from London to New Towns and growth areas in the South East and to Assisted Areas". (Royal County of Berkshire 1976, p.77). In the early years of the policy, there were high levels of refusal of ODPs, as indicated in Table 3.1.

Table 3.1: ODPs in Central Berkshire: Approved and Refused Floorspace

	ODP Approvals		ODP Refusals		
	No. of Schemes	Sq. Metres	No. of Schemes	Sq. Metres	% (Floorspace) Refused
1965-66	13	9,200	7	35,580	79.4
1967-68	29	40,780	11	27,590	40.4
1969-70	14	53,140	5	32,700	38.1
1971-72	19	153,750	12	110,650	41.8
1973-74	19	116,130	9	38,460	24.9
Total 1965-74	94	372,990	44	244,980	39.6

Source: Adapted from Royal County of Berkshire Structure Plan, Report of Survey 1977.

By the early 1970s, it will be noted the policy was becoming less restrictive and the increased proportion of approvals reflects the encouragement for office development in growth areas designated in the Strategic Plan.

In total, between January 1965 and December 1974, ODPs were granted for 373,000 square metres of office floorspace in the area as a whole - 69% for Reading and 28% for the new town of Bracknell- The general relaxation of policy towards office development in the outer south east, contributed to very rapid growth during the 1970s. By 1982, Reading had 214,000 square metres of office space, and was one of the most sought after office locations outside London (see Chapter Five). The Strategic Plan undoubtedly assisted the outer areas of the

South East as illustrated by the fact that in the two years after its publication, between July 1970 and April 1972, ODPs were issues for a net area of 1.022 million square metres of offices in growth areas and new or expanding towns in the South East, with just less than 465,000 square metres refused permission. Elsewhere in the outer metropolitan area and the outer South East, only 557,000 square metres were permitted, with a similar amount refused.

What effect did the control system imposed at a national level have on the development process? The immediate impact in London was to help to push rents up to unprecedented levels, placing a considerable scarcity value on office space in the centre of the capital. Alexander cites a rise from typical levels of £2 per square foot per annum in 1963, to £5 by 1969 and £14/£18 by the early 1970s. (Alexander 1979). Such rises obviously placed very high values on properties and existing owners were sitting on assets which grew in value almost daily as a result of government actions.

It was equally evident that the suppliers of office space, the investment institutions were selective in their activity. This was a natural result of their motives for investing in office property since only developments with an assurance of security and financial viability would be encouraged. Obviously such developers were most liable to be found in areas of high demand for offices and not in the more peripheral regions. The outcome of that logic was for London and the South East region to be the most favoured recipients of the property investment finance of major institutions. Indeed Colenutt and Hamnett have shown that in a typical pension fund's portfolio, the geographical distribution of office investment was by 1979 still very much dominated by London and the South East as illustrated by Table 3.2.

Table 3.2: Mineworkers' Pension Fund. Office Property Holdings (excluding combined shop/office developments) (£m).

			% Total
London	Central	30.5	26.1%
	Outer	34	29.1
England	NE	4	3.4
	NW	8.5	7.3
	Midlands	0.5	0.4
	East Anglia	4.0	3.4
	SE	18.5	15.8
	SW	10	8.5

(cont)

Table 3.2 (cont'd).

		% Total
Scotland	1.5	1.3
Wales	4	3.4
Ireland N	1.5	1.3
UK Total	117.0 m.	(36.2% of total property portfolio)

Source: Adapted from Figure 15b, of Unit 9, Urban Land Use and the Property Development Industry, Urban Change and Conflict, Open University.

A shift in policy planning concern office location is evident in the case of the structure plan of another growth area in south-east England. South Hampshire, designed as an area of potential growth since the South East Study (HMSO 1961), and confirmed in that role in successive regional planning statements from that time, produced its structure plan in draft form in 1972 (Hampshire County Council 1972). In it, the forecast of growth in the office sector suggested a rise from 53,500 detached office jobs in 1960, to 79,900 in 1981 and 99,500 in 1991, requiring some 330,600 square metres of new space by 1991 with an additional 186,000 square metres of space to replace obsolete office buildings. The policies contained in the plan emphasised the continuing role of the centres of Southampton and Portsmouth as office locations, although some provision was made for new office development in suburban centres and the need for campus office developments for prestige office headquarters was also recognised. The examination of the structure plan by the Department of the Environment produced some modifications, including the direction of a larger proportion of new office employment to suburban centres when the final plan was approved in 1976.

By 1982 and the up-dating of the structure plan, provision was made for 25,000 square metres per annum of office development up to 1996, with a total of 45,000 square metres allocated to the two city centres, 10,000 square metres each in the three suburban centres of Eastleigh, Fareham and Havant/Waterlooville and 16,500 square metres on each of two campus sites. The provision of office space was therefore fairly closely circumscribed, but it is interesting to note the move towards planning specifically for suburban growth. In addition, there is specific mention of the need to control speculative office development. (Hampshire County Council 1982, p.21).

In practice, the planning policies have tended to reflect the demands of the market, and as such have gradually moved away from an over-concentration at the city centre to permit more peripheral developments. This can of course be justified in terms of decreasing journeys to work and traffic congestion, but it is also true that the development industry favoured out-of-town developments. An example is provided by the outline approval granted in January 1984 for a 60 hectare business park adjacent to the M27 between Southampton and Portsmouth. The Solent Business Park is to be developed by a company largely comprised of Arlington Securities and Philip Hill Investment Trust, involving an investment of £100 million. (Financial Times, 13 January 1984). Any mix of office and production space is to be permitted, although there will be a restriction on the amount of warehousing in the development.

It is evident, that even in the UK, where planning controls have been relatively strong, the investment market eventually dictates the pattern and direction of development. Certainly the evidence in the UK is that in the long term at least, it is difficult to totally constrain the office development process. Only rarely have local authorities been willing to impose strong planning controls to preclude office development. The case of inner London will be discussed further in Chapter Five, but it is worth noting in this context that although there are very strong policies against offices in some London boroughs, such policies are often regarded as being rooted in left-wing political judgements. More usually, politicians of any persuasion and the planning authorities which they control, have been reluctant to turn down investment capital and a local income return and planning policies have been eventually shaped to accommodate the development process.

CONTROL OF INVESTMENT FINANCE

Our discussion thus far has concentrated on the UK as an example of widespread planning controls on the office development process. Many other countries of course have imposed controls and some of these will emerge in later chapters (see also Alexander 1979, pp.62-98, for further examples of control policies). In the context of this study, however, it is appropriate to turn to other forms of control rarely discussed elsewhere which have a marked impact on office development. As Chapter Two indicated, the flow of investment overseas into property is a vital stimulant to office development activity. Not only British finance is involved, since investment flows from amongst others, Germany, Netherlands, France, Canada, the USA and countries of the Far East into other countries to promote property development. Many examples of such movements of capital will emerge in later chapters, underlining the importance of this international flow of finance. The cross-currents of international investment are not however, uncontrolled, since there are often obstacles to international investment which can be erected either by the exporting or the receiving country, or by both.

Foreign ownership of land is often precluded by legislation or made to satisfy certain conditions. Several European countries have controls over such ownership. The 1917 and 1974 Concession laws in Norway, for instance, meant that foreign nationals required consents for the acquisition of freeholds or leaseholds of more than ten years' duration. Most countries require overseas investors to seek permission prior to large-scale investment, although some countries, such as Belgium are rather freer than others in this respect. In some cases, the purpose of the investment may well determine whether such permission is required. In the case of Norway, cited above, consents are more likely to be obtained if it can be demonstrated that production and employment is being created. Speculative property development may well be less favoured.

On the other hand, much of the legislation passed by government has been designed to limit foreign ownership of resources. In Sweden for instance, British ownership of the iron ore industry early in this century, led to legislation against alien ownership, which is still most likely to prevent ownership of forestry and of mineral rights. Similar concerns led to the passing of the Foreign Investment Review Act in Canada in 1973, which controls both the acquisition of businesses by foreigners and the setting up of new ones. A requirement of the Act is that foreign investment must be demonstrated to be of positive benefit to Canada rather than merely not being detrimental. Thus the Foreign Investment Review Agency recommends to the Federal Cabinet which business enterprises - which includes real estate valued in excess of C$10 million - should be permitted.

There is a similar restriction in Australia, where the Foreign Investment Review Board must approve every real estate investment over A$350,000, equity participation must involve a minimum 50% Australian interests and furthermore, there must be proven net economic benefits to the Australian economy. Such obstacles may well be difficult to surmount and may involve some intricate financial dealings, such as the British Property Company, MEPC's acquisition of a minority share in a property unit trust in order to revitalise a property development in Australia (Financial Times, 10 February 1984).

Domestic controls over investment in property by financial institutions are further restraints on the geographical distribution of investment capital. In Britain, there has been a general tendency since 1979 for restrictions to be lifted to permit a greater spread of capital overseas. The lifting of exchange controls in 1979 was a major measure which released a flow of capital abroad, both into property and into equities. Net investment in property overseas particularly demonstrates this, as does total investment by British interests overseas, shown in Table 3.3.

Table 3.3: Property and Total Investments Overseas by British Companies 1977-81 (£m)

	Property	Total
1977	-17.2	1884.8
1978	-14.6	2709.7
1979	29.6	2776.7
1980	376.6	3430.0
1981	164.9	5103.6

Source: Business Monitor M4 1981 Overseas Transactions (HMSO 1983).

It should of course be noted that other factors such as the recovery of the property market from the mid-1970s also had their effect on these figures, but the trend is nonetheless well marked.

Until 1983, pension funds of local authorities in Britain were limited in their ability to invest overseas. The previous regulations stipulated a minimum holding of 25% of a local authority fund's portfolio in fixed interest investments and a maximum of 25% in overseas and property investments combined. Many funds had invested their maximum in property when the restriction was lifted with effect from October 1983. One example is the Greater Manchester County Superannuation Fund which held 5% of its funds in overseas investments and 20% in property. In 1982-83, the Fund worth £680 million, invested only £4.4 million in property, presumably because its holdings were close to the legal limits. (Estates Gazette, 8th October 1983). The removal of such restrictions, however, may not have had the effect of stimulating new investment in property since as has been commented upon in Chapter One, returns on property in 1983 were not especially attractive. On the other hand, the freedom to invest in overseas industrial equities seemed likely to promote the outward flow of capital.

Capital flows into the office development process are therefore not always free and the degree of control may vary from time to time and between places. In Western Europe both Switzerland and Germany place restrictions on their investing institutions to limit the outward flow of investment capital. On an international scale, these limitations on the development process may be significant in channeling funds to particular locations and away from others.

Ownership of land may be a further factor which acts as a control on development. In the case of Hong Kong, discussed in the following chapter, the colony's government has been very active as a stimulant to office development by releasing land, often reclaimed, for development. At the same time, it has been able to regulate the pace of development by programming the release of development land. In the recent past, however, its high dependence on the revenue so earned and a declining property market has markedly diminished the government's ability to control the release of land for development.

The office development process operates therefore in a series of structural constraints. At the local level of structure and local planning, it must conform to policy guidelines, although the interests in property are sufficiently powerful for them to begin to shape policy in some areas. At the international level, restrictions on capital flows act as a break on investment finance being used to promote office development. It must be said, however, that there are few examples of controls so complete as to prohibit totally foreign investment. The following chapter examines the consequences of the free flow of such investment, whilst in succeeding chapters, a spectrum of *laissez-faire* to strong control at the local level emerges from the case studies and cities considered.

Chapter Four

PHYSICAL IMPACT OF THE DEVELOPMENT PROCESS

The activities of the property market, more or less constrained by planning controls, find a concrete expression in the city itself and more specifically, in its built form. Studies of urban morphology have identified cycles of development activity, (see for instance, Conzen (1960) and Whitehand (1972)). Yet little close investigation exists of the processes of change in the city as it adjusts often over a short period of time to the cyclical pattern of redevelopment. Early studies of the central area recognised its dynamics, particularly its extension and contraction (Murphy and Vance (1954) and Horwood and Boyce (1959)). Through these studies emerged the concepts of zones of discard and assimilation, or areas from which functions were withdrawing as opposed to those into which new developments and their functions were moving Modern office development, however, causes changes both within the existing central area and to a lesser extent on its periphery in a process of spatial extension. Studies of central areas of urban centres in West Yorkshire have demonstrated the type of change which took place during the early post-war property "boom" of the late 1950s and early 1960s (Bateman 1971). J. W. R. Whitehand has recently reviewed the nature and management of change in city centres pointing to the "large gap between the way in which British city-centre townscapes appear to be evolving in the 1980s and our suggestions as to the principles which should govern the management of change." (Whitehand 1983, p.57).

SPATIAL IMPACT OF THE OFFICE DEVELOPMENT PROCESS

It would be false to conclude that only the city centre has been affected by the office property development cycle since, particularly in North America, a pattern of new office developments can be traced all the way from the city centre through the expanding suburb to the office park on the periphery of the city. In some cities, recent transport develop-

DOI: 10.4324/9781003174622-4

ments have assisted this process endorsing a new pattern of accessibility at new transport interchanges which have then been developed by property interests to produce new office modes. Successive interchanges at subway stations northwards along Yonge Street, Toronto are a good example, whilst the building of the Mass Transit System in Hong Kong has offered similar opportunities, to be discussed in detail below.

In many cities, for historical reasons, there is a favoured direction of expansion which guides the pattern of office and other commercial development. It is perpetuated because few property developers are willing to be 'trail-blazers' into the uncharted territory of unfashionable zones of city, where their investment would be unsure of the insurance of universal acceptance of its value. On the other hand, a favoured direction of growth can occur if one or two prestige investors invest in an area since it may rapidly acquire both fashionable status and the investment funds of other institutions.

The peculiarly mutual interdependence of the financing institutions discussed in the previous chapter is further reason for them to be mutually supportive in terms of favouring one area more than another in a city. The process becomes self-perpetuating as land values rise, followed by an increase in rents which can only be met by prestige clients who in turn attract further investment into the vicinity to begin the cycle once again. In most cities, however, it would be jeopardising the financial position of an investing company to ignore the realities of the local market. In a few, nonetheless, such as Paris, London and New York, the market is sufficiently differentiated in a spatial sense to permit profitable investment in so-called secondary locations away from the most sought-after areas of the city.

A similar pattern of common decision-making leading to the emergence of popular areas for office investment exists at the regional scale. For instance the outward movement of investment along the corridor containing the M4 motorway west of London means that certain sites are favoured at the possible expense of others. Thus Bernard Thorpe's Annual Property Survey 1983 showed that rents in the sector west of London included high values for Chiswick, Ealing and Hammersmith (£10-£13 per square foot), extending to a level of £12 per square foot in Reading, the highest rent level outside London. Whilst Bristol and Cardiff offices could not command these rents, levels of £6 and £5 per square foot respectively were respectable provincial city rents comparable with larger provincial cities such as Birmingham (£4-£7.50) and Manchester (£6.50). The popularity of Reading for office users and thus office developers is certainly attributable for its communications advantages but it is also partially a product of the communal perception of decision-makers.

TEMPORAL DIMENSION OF THE OFFICE DEVELOPMENT PROCESS

Office development is a cyclical process and thus affects a city for a limited period of time. The length of the cycle may vary from both city to city and one county to another. There are a number of reasons for this. One city may be consistently seen as a sound investment by those responsible for deciding on the location of property finance. London and Paris both have enjoyed this relative permanence. Others, however, may have only a brief period during which investment finance is directed towards them. Poor returns, threats of control or political events may each determine that the flow of investment finance is directed elsewhere. There is little evidence that the cycle of office development follows the normally smooth curve of a development cycle. Instead, whilst its upturn may well be relatively normal, the downturn of the cycle may be quite precipitous occasioned by a sudden loss of confidence in that particular office market. In both detailed case studies in the remainder of this chapter, the office development cycle will be seen to have had remarkably rapid terminations.

CASE STUDIES OF PHYSICAL IMPACT OF OFFICE DEVELOPMENT

Two case studies of office development will suffice to demonstrate the effect of large scale investment in offices. Both Brussels and Hong Kong enjoyed the confidence of property investors since each was seen as cities in which offices would be in continual demand. In the case of Brussels, its status as a major European capital endorsed it with an added attraction. Hong Kong, on the other hand, with its history as a trading colony developed a thriving industrial structure, temporarily keeping ahead of the world recession of the late 1970s and early 1980s with a considerable influx of foreign companies to enhance its commercial importance. Both cities saw a cycle of development during which property developers transformed large tracts of the city, only for the cycle to terminate leaving a legacy of large-scale office over-supply. There are, of course, important differences between the two cities. Brussels saw development primarily within its central area or its extensions. Hong Kong's office development certainly affected its traditional CBD but spread also into a number of important decentralised locations. The promotors of developments also varied since in Brussels, British property developers were extremely active whereas in Hong Kong, a mixture of south-east Asian finance was eventually joined by British property interests to fuel a much more short-lived development cycle which lasted only four years.

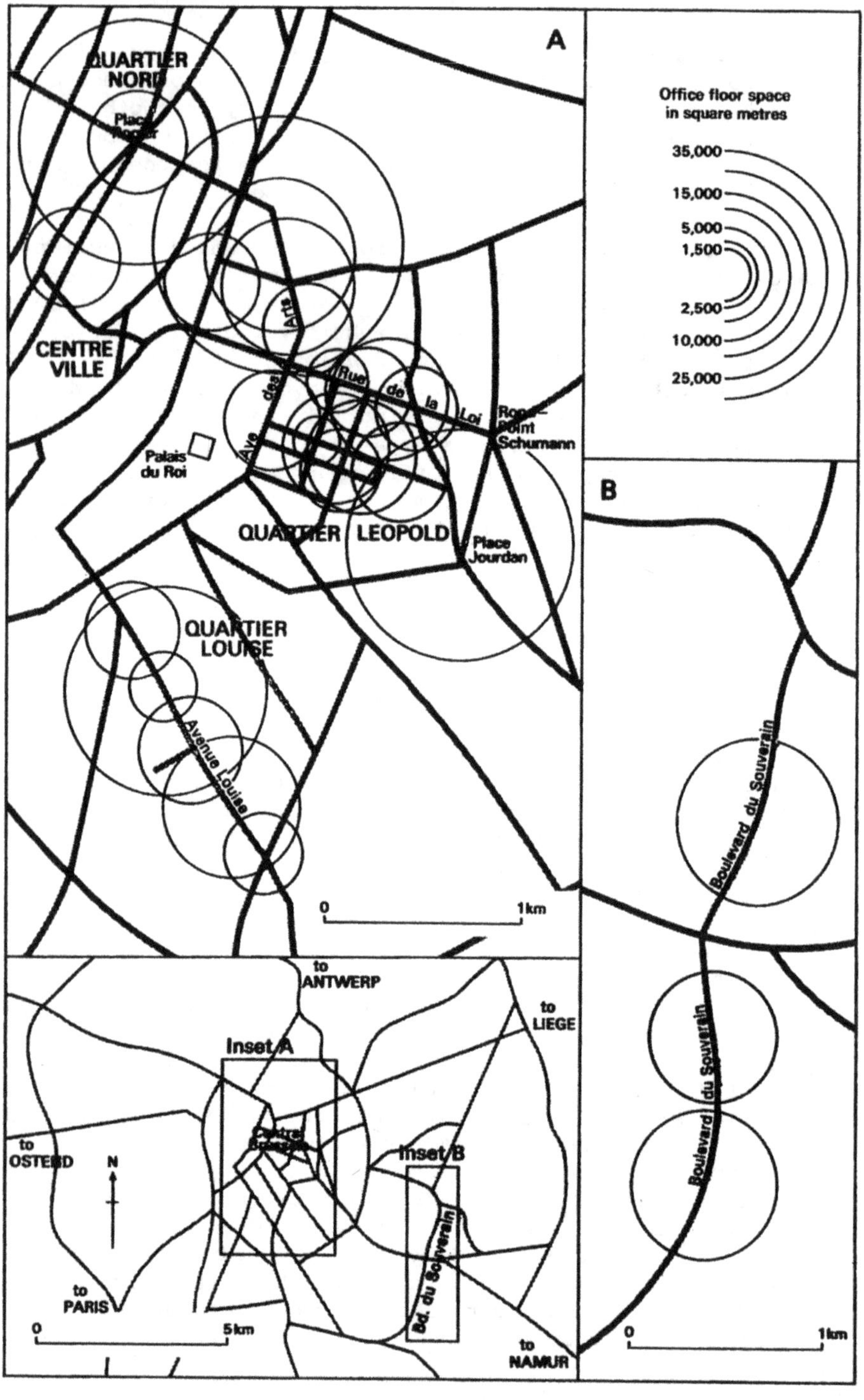

Fig. 4.1 British Investment in Offices in Central Brussels 1972 - 1973

Brussels

By the early 1970s, Brussels had emerged as the embryonic European capital. Besides being in its own right a national capital, it housed the headquarters of NATO and the EEC. It was attractive to international companies which required a base in Europe, as well as to foreign governments, many of whom wished to maintain a "listening post" close to the economic heart of European commerce. At that time, British property developers were seeking alternative investments, since the returns on property in London were no longer easily available in the face of increasing office development control. Within the city, certain areas were especially favoured by developers with the Quartier Léopold and the Avenue des Arts being particularly important foci of the property developers attentions. It was also evident that other newer areas were emerging especially the Manhattan World Trade Centre complex to the north of the city centre. Firmly established, however, was the area to the east of the historic centre, with the principal areas of Avenue des Arts, Avenue Louise, Rue de la Loi, which linked the traditional central area to the principal EEC offices.

The location of British property companies' developments in the period from 1972-1973 confirmed this concentration of activity. (See Figure 4.1). Table 4.1 indicates the intensity of British interest and the number of companies involved.

By 1974, some caution was being exercised. A downturn in the cycle of development was clearly taking place reflecting the recession induced by the oil crisis of 1973-1974 and property companies had to respond to it, resulting in the withdrawal of some property companies. There were also new restrictions by the Belgian government who had at last awakened to the need to control the spiralling office development in Brussels. It imposed what amounted to a twelve month embargo on all commercial building costing more than FrB50 million, effectively more than 25,000 square feet (2,322 square metres) which still lacked final planning permission. Banks were also restricted in their ability to lend finance for property and all mortgage bank lending ceased.

But the supply situation was not controlled and certainly not in balance with demand. In 1975, 3.92 million square feet (375,884 square metres) of office space became available, with a further 4.42 million square feet (423,828 square metres) scheduled for the following year. With an annual take-up of 2.3 million square feet (220,544 square metres) an oversupply situation existed which was to be a continuing problem into the 1980s. Table 4.2 showing the range of rent levels for good office accommodation clearly indicates the manner in which the supply was in excess of demand from 1973 onwards and rents dropped from that date. They failed to retain their 1973 peak until 1982, and even then, not in real

Table 4.1: British Property Companies' Office Developments reported in Brussels. 1972-73 (excluding acquisitions of completed property)

Company	Location	Floor Space (sq. metres)
C. H. Beazer (Holdings) Ltd.	Ave. des Arts	4,794
Amalgamated Investments and Property Co.Ltd.	Ave. des Arts	5,202
Brixton Estates Ltd.	Ave. Louise	11,148
Town and City Properties	Rue de la Loi	3,865
Heron Corporation	Ave. Louise	3,716
Grosvenor Square Property Co.	Ave. Louise	251
Artagen Properties	(off) Ave. des Arts	348
Reamhurst Properties	Bd. du Souverain	15,734
Reamhurst Properties	Ave. Louise	6,500
Grasshopper Unit Trust	Ave. Galilee	3,530
Imry Property Holdings	Rue de la Loi	418
St. Martin's Property Corporation*	Rue Royale (Centre Nord)	511
Artagen Properties	Rue Montoyer (Q. Leopold)	643
Town and Commercial Properties	Ave. Louise	24,990
Stead Investments Belgium S.A.	Bd. du Souverain	10.196
Abbey Property Bond Fund	Bd. du Souverain	13,006
Property Security Investment Trust	(off) Ave. des Arts	7,432
Rank City Wall	Pl. de la Monnaie (Centre Ville)	530
Ernest Ireland Property Europea S.A.	Rue Belliard (Q. Leopold)	5,946
	Rue de la Loi	2,206
Irish Life Assurance Co. (and Hardwicke)	Rue de la Loi	780
Estates Property Investment Co.	Rue Froissart (Q. Leopold)	2,787
Star (GB) Holdings	Bd. Bischoffsheim (Ave. des Arts)	13,470
	Bd. Bischoffsheim (Ave. des Arts)	10,683
	Bd. Jardin Botanique (Centre Nord)	6,038
Commercial Union Properties	Ave. des Arts	18,580

*Kuwaiti-owned

Table 4.2: Brussels Office Rents 1970-1981

	Range - Belgian francs per square metre
1970	2,250-2,500
1971	2,250-2,750
1972	2,750-3,250
1973	3,000-3,750
1974	3,000-3,500
1975	2,750-3,500
1976	2,500-3,000
1977	2,500-3,000
1978	2,750-3,000
1979	2,750-3,500
1980	3,000-3,750
1981	3,000-3,900

Source: Financial Times, 4 June 1982.

terms, although the larger demand for offices commanding the highest rents was noticeable at the end of this period.

Some significant developments affecting the office market in the city took place in 1975. Many British companies were at that time being affected by the downturn in the British property market. For many companies overseas schemes which were poorly financed had to be postponed or abandoned particularly if the finance was secured against British property. At the same time, some schemes were being completed, adding to the growing stock of offices available in the city, such as the Commercial Union Properties' 15,500 square metre development on the Avenue des Arts and Artagen Properties development, in the centre of the Quartier Léopold, of 6,180 square metres of offices at the junction of Rue Montoyer and Rue d'Arlon.

The major British agents were contrasted in their views concerning the direction of the property cycle. Jones Lang Wootton were optimistic, pointing to a letting of 29,645 square metres in the first four months of 1975, with leases agreed on a further 21,028 square metres. Most of the offices let were in the Quartier Léopold, the traditional city centre and the Quartier Louise. (Estates Gazette, 24 May 1975). Richard Ellis, on the other hand, predicted an oversupply of 245,000 square metres by 1976 (Estates Gazette, 31 May 1975).

By this time, property agents Richard Ellis recognised six distinct office quarters in the city, besides decentralised locations. (Estates Gazette 31 May 1975). The quarters were:

1. <u>Avenue des Arts/Boulevard du Regent</u> - still the most prestigious location, commanding the highest rent levels in the city.

2. Quartier Leopold - further away from the centre, but favoured in 1975 for long term investment prospects.
3. Quartier Louise - an area focused on the Avenue Louise which had been particularly popular with international companies seeking office accommodation in the city.
4. Centre Ville - an area of very much more restricted growth of new offices.
5. Quartier Nord - one of the early expansions of the central commercial zone to the north of the city centre, with office developments dating from the 1960s.
6. Quartier Schumann - an emerging area centred on the EEC headquarters.

A series of major developments in the suburbs were also in progress, the most important location being the Boulevard du Souverain. British companies were active in that location with the commencement of a 15,800 square metre project called Souverain 100 by Reamhurst Properties and the completion of Abbey Life's Parc de Seny project providing 13,000 square metres of office space in landscaped surroundings.

The property industry had in a short period of time, made a considerable impact on the city. Areas to the east of the city centre were particularly affected since most of the defined office quarters were in that area (see Figure 4.1).

The development process did not continue totally without hindrance, however, since certain legislative measures were taken in 1975 in acknowledgement of the need to restrain the activities of the developers. The plans for the Avenue Louise were altered in an attempt to safeguard the area from further large-scale alteration, although it should be noted that the area had already been the subject of a royal decree controlling new development in 1972. Central government announced the suspension of building permits for the centre of the city pending the revision of the plan for the city centre. But more significant for office developers than attempts at planning control were a number of economic measures passed in November 1975 by the Belgian government (Estates Gazette, 8 November 1975).

Firstly, as part of a series of anti-inflation measures, rents were frozen initially until the end of 1976. Secondly, the local authority for the Brussels agglomeration (Conseil d'Agglomeration) imposed a development tax on office building. It was to be retro-actively applied to all office buildings begun after January 1974, and to last until the end of 1976. Initially a level of BF350 per square metre was suggested, but it was finally set at BF98F per square metre. Speculative offices were particularly penalised since the developers of owner-occupied premises could claim a 33% reduction in the level of the tax. Those who were developing offices in Brussels merely to make a profit rapidly, were also more heavily penalised since developers who were carrying out residential development elsewhere in the agglomeration could claim a 90% remission of tax on a *quid pro quo* basis, offsetting each square metre of an office scheme against equivalent residential development. Further controls were placed for the first time by central government on the conversion of residential properties to offices. The government also passed legislation to curb the practice of allowing buildings of historic importance to deteriorate beyond repair, with the intention of thereby justifying permits for demolition.

To summarise, therefore, by 1976 the situation was weak for office developers. New controls were in operation and restrictions on rents frequently made even existing properties unprofitable. Some of the hitherto favoured areas for development were beginning to show worrying signs of weakness for developers. For instance, the Avenue Louise area, favoured until 1975 was suffering on a number of counts. Traffic congestion was one unforeseen result of the vast increase in offices. Furthermore, the new Brussels métro which opened for service in 1976 did little to assist since it served Avenue de la Loi and the Quartier Schumann, but not the Quartier Louise. (Financial Times, 20 October 1976). Estates Gazette commented in December 1976 that, "according to the agents (St. Quintin Son and Stanley), Avenue Louise still fails to realise its ambitions, and what was once one of the more attractive residential avenues is now over characterised by serious over-supply of offices, exacerbated by poor public transport and restricted parking facilities." (Estates Gazette, 18/25 December 1976). In all areas of Brussels, however, developers were being forced into the position of selling investments often at prices below the then current development costs, since rent control eroded their potential profits.

Whilst activity in the office market did not cease entirely, it was certainly more restricted from 1977 onwards. It was estimated that approximately 500,000 square metres of offices were in surplus in the city, (de Wandeleer, 1977). Rents once again reflected this, even in the Quartier Léopold where rents in early 1977 were in the range of BF2750-3000 per square metre, compared to BF3250-3750 per square metre in the preceding year. The control on rents was extended into 1977, with a freeze on rents for all leases granted after 1st April 1976, whilst earlier leases whose rents were already frozen were permitted to rise 8.5% which was close to the 9.3% increase which would have been achieved by indexation (Estates Gazette, 12 December 1977). It should be noted that there was still some demand for office space in the city. The growth of the EEC and the Belgian government's reluctance to decentralise ensured that office jobs were still rising at a rate of 10,000 new employees in this section each year requiring 75,000 square metres of offices. (Vickers da Costa, Quarterly

Property Review, February 1977). A supply of at least 500,000 square metres, however, underlined the surplus and the problems for property developers. Trading in office blocks still took place as companies attempted to balance their assets and adjust to the new market conditions. One of the most significant sales was that of the Tour Astro on the Avenue des Arts. This building contained some 35,000 square metres of office space and was begun at the height of the office boom by Barclays Bank Trust Company and London and Overseas International BV on behalf of the Grasshopper Property Unit Trust and was completed in 1976. By October 1977 only a small area had been let of the 32-storey building and it was sold to a local semi-official body, *la Société Nationale de Crédit a l'Industrie*, which itself occupied just over 50% of the available space. By that time because of the glut of office space, and the fall in the value of sterling against the Belgian franc, the value of the building had in fact fallen below its cost. (Estates Gazette, 29 October 1977). The Tour Astro sale was only one of several sales of completed buildings during 1977, another example being an 11,206 square metre office block in the Quartier Léopold and close to Rond-Point Schumann developed by Slater Walker and sold to the Shell Pension Fund for £8 million in late 1977, when it was let to the EEC. The significance of this sale is that it pointed the way to a tendency for funding to come increasingly directly from the institutions rather than through traditional developers simply because they were more able to await an upturn in the property market. The more direct participation of the institutions in property dealings was becoming more common in all property markets, as discussed in Chapter 2.

The period from 1978-1980 was characterised by a similar pattern of some selling activity and some adjustment to the legislation affecting property. A new property tax was proposed, but not implemented, whilst rent increases were once again controlled so that although they could be indexed to inflation, this could not happen until after one year's occupation of the premises.

By the beginning of 1980, Jones Lang Wootton estimated that of the stock of 5.75 million square metres (60 million square feet) of purpose-built offices in the city, 239,722 square metres (2.5 million square feet), or 4.2% were vacant. (Estates Gazette, 22 March 1980). The continuing expansion of the EEC was seen as a good sign and Jones Lang Wootton itself had let 153,422 square metres (1.6 million square feet) of offices during 1975. In the period from 1976-1980, the British agents had let 153,422 square metres (1.6 million square feet) of offices to the Belgian government, 57,533 square metres (600,000 square feet) to foreign administrations and 148,628 square metres (1.55 million square feet) to foreign companies, indicating the extent to which the office market in Brussels was becoming increasingly dependent on demand from the public sector.

The 1980s saw no respite, despite earlier optimism by the major letting agents. Economic and political instability inhibited further investment. In addition, the practice of indexing wage increases to inflation led to very high wage costs nationally. Certainly some international companies elected to leave Brussels altogether. For instance, the Sheraton Management Corporation, managing its international hotel interests, moved its European base to the UK at Denham Place, Buckinghamshire (Financial Times, 6 June 1980).

By early 1981, some rent rises were taking place, making new development once again a possibility for traditional property developers. It should be noted however, that it was felt at that time that rent levels needed to double to ensure profitable investment by developers dependent on external credit. The period 1980-1981 saw some other government measures which affected the office market in Belgium more generally. A limited degree of devolution to regional assemblies in Flanders and Wallonia, which had always been a fraught issue in the bi-lingual state, took place. This led to an increase in office demand in some centres such as Ghent, Bruges, Hasselt, Liège, Charleroi and Namur. (Estate Times Review, January 1981). Antwerp was particularly affected with rents for modern office space in the city reaching and exceeding those for central Brussels. (Investors Chronicle, 3 April 1981).

In a review in mid-1981, the Financial Times chronicled the continuing malaise of the Brussels office market. Dutch pension fund investment which had accompanied British investment in the 1970s was, like its British counterpart, increasingly being channeled to North American property. The Dutch development company Wereldhave was quoted as having seen its investment in Belgian property fall from 6% to 5% of its property portfolio between 1979 and the end of 1980, whilst funds invested in the USA rose from 18% to 21%. (Financial Times, 15 May 1981).

Some development activity was continuing, but in the face of a continuing over-supply. Slough Estates leased its 17,547 square metre (183,000 square feet) Léopold Business Centre to the EEC in October 1981, but the building had been lying empty for five years prior to that. Some long term investments were also made such as the purchase by the Kuwaiti government's St. Martin's Property Corporation of a 16,780 square metre (175,000 square feet) office development in Rue de la Roi, at its junction with the Avenue des Arts.

A joint survey by Jones Lang Wootton and l'Université Catholique de Louvain (1980) reviewed the position of the office market after the activity of the 1970s. It saw a need for an annual supply of 150,000 square metres of office space. In fact the average annual supply from 1970-1980 was

170,000 square metres. The availability of new buildings was 150,000 square metres whilst the second-hand building stock was 286,000 square metres. With 290,000 square metres under construction of which over 65,500 square metres was speculative, and a further 175,000 square metres approved, it was clear that there would be a continuing surplus of office space. The survey also pointed to the potential for new office development in the west of the city, to be served by an expanded métro system. The report was optimistic concerning the continuing European importance of Brussels and noted that little government office work had in fact left Brussels, although the private sector had stimulated office activity in regional centres, in anticipation of government decentralisation.

In 1982, there were continuing problems which prevented the long awaited upturn in the Brussels office market. In particular, economic and political problems persisted, resulting in devaluation of the Belgian franc in February and an unemployment rate in March of 14.7%, a state of affairs which did little to reassure the overseas commercial interests who were needed to revitalise the office sector. On the other hand, the lack of any significant building programme, resulting from the poor market conditions of the preceding years did mean that at least the oversupply situation was not being exacerbated and indeed there was some feeling that there was a shortage of good, large office units in the principal locations. (Financial Times, 16 April 1982). Consequently rents in these areas did re-attain their 1973 peak levels approaching BF4000 per square metre, although admittedly not in real terms. The situation was still not particularly encouraging however, especially as rents of BF5000 per square metre were needed to stimulate the financing of new development. Vacancy levels were estimated at 57,000 square metres by May 1982 (Estates Gazette, 8 May 1982), a 60% fall from the preceding year. One result of this, however, was to leave a growing surplus of second-hand office space as tenants moved to more modern premises, creating a new market for refurbished office accommodation.

The market by 1983 had confirmed the trend of decreasing vacancy rates, with a figure of 4.6% quoted for mid-1983 (Financial Times, 9 September 1983). A symptom of the shortage of space was that the EEC had acquired a building of 4,600 square metres, in contrast to its usual preference for larger office units, which were not available at that time. The lack of confidence in the early 1980s had meant that large new projects were not becoming available in 1983.

The saga of Brussels office development is important for a number of reasons. It illustrates well the effect of short-term changes in market conditions on the office supply. The almost uncontrolled activities of property developers exaggerated the impact of the economic cycles which affected the whole of the western world economy in the 1970s and early

1980s. The tangible effect on the city was considerable as certain parts of it were transformed by new buildings and functions with vastly increased building densities. It was hardly surprising that local community resentment against office development was rife, although it took time for effective response to come from either central or local government.

Hong Kong

A second case study of dramatic physical impact arising from the activities of the property market is presented by Hong Kong. In this case, a number of factors combined to produce a remarkable cycle of rapid growth and equally rapid collapse from which recovery has only recently begun. The cycle of development concerned is very much shorter than even that described above in Brussels and certainly far shorter than more traditionally recognised building cycles. From a low point stemming from the virtual collapse of the Hong Kong stock market in 1973 when the Hang Seng index fell from 1700 to a low of 160.42 in 1975, there was a gradual increase in property development activity including that of offices, which accelerated in 1979 and continued to gather pace until 1982. The finance came from British property companies and more traditional sources elsewhere in South East Asia. The cycle of growth was interrupted by political uncertainty stemming from Hong Kong's leasehold status. Whilst this was the trigger mechanism, however, other factors combined to undermine the commercial confidence so important to the maintenance of the value of property. It could be argued that faith in the inherent value of land, discussed in Chapter Two, had by any measure, been pursued to an unreal level, best illustrated by the sale of a piece of land for development in 1982 at a price of £346 million per hectare.

Hong Kong presents, therefore, some very different conditions from those prevailing in Brussels, but is worthy of our attention simply because of the scale of physical transformation and the cyclical nature of the process. An understanding of its political background is essential to a review of its property activities.

The mainland New Territories and parts of Kowloon are leased from China until 1997, whilst the original trading colony, including the present central business district of Hong Kong Island, together with parts of Kowloon peninsula were ceded to the British crown to be held in perpetuity. As the expiry of the lease approaches, political negotiations have affected the development process, *via* the property market, despite the fact that the eventual status of the leased territory is by no means clear. Property investment has included developments in both parts of the colony and has already included some activities by the Chinese government itself in acquiring land in the New Territories.

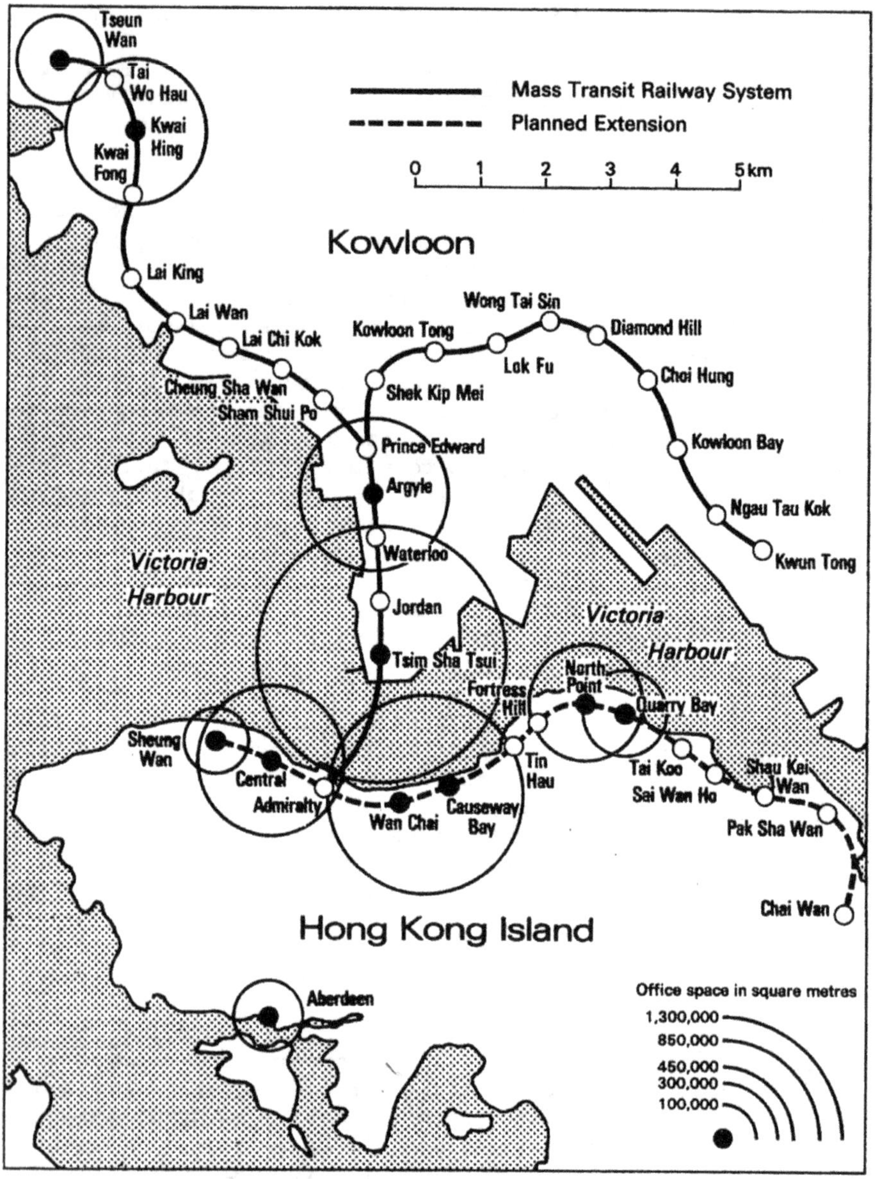

Fig. 4.2 Hong Kong Office Pattern 1980 - 1984

Physical Impact of the Development Process

In theory at least, control of the development of Hong Kong is in the hands of the colony's government, since all land for development is owned by the government, including reclaimed land. The government releases land for development, on a leasehold basis, with leases in theory up to 999 years, but in practice 75 years, or much shorter in those areas affected by expiry of the Crown lease. The leases are sold to the highest bidder and this revenue has been of considerable importance to the colony, yielding in 1978-1979, 15% of total government revenues, rising in 1980-1981 to 35%. Obviously such dependence on land sales brings with it considerable vulnerability when land prices fall as they began to in 1982. It led in 1982-1983 to a budget deficit for the colony, since of the HK$12 billion (approximately £1.08 billion) expected from the sale of land to developers, amounting to 30% of the budget revenue, only 50% was actually realised (Financial Times, 11 November 1982). Such shortfalls in turn lead to the situation in which more development land needs to be sold to raise revenues, but that in itself is self-defeating in a declining property market. As an example of the basic underlying weaknesses of western capitalism, as played out in the artificial world of the property market, Hong Kong could hardly be emulated.

A further factor important to the development process was the decision in the early 1970s to build an underground railway system to solve the chronic transport problems of the colony. As planned, the Mass Transit Railway (MTR) was a 20 kilometre subway system with twenty stations, linking Hong Kong with the Kowloon peninsula. The building of the MTR provided the opportunity for decentralised office locations to be developed away from the traditional CBD. These centres began to be important in adding to the office stock of the colony during the period 1980-1982 (see Figure 4.2).

There has been in Hong Kong, a number of very prominent property companies whose activities have frequently shaped the general pattern and conditions of the market. The interlinkages between companies and trading interests in the colony are labrythine, and the early British merchant companies are still well represented. Thus the largest property development company, Hong Kong Land has very close ties to the British merchant banking and trading concern of Jardine Mathieson. In its turn, Hong Kong Land has not been involved exclusively in property, but also in public utilities, holding a one third equity stake in the Hong Kong Telephone Company, sold in March 1983 for HK$1.4 billion, and a one third ownership of the Hong Kong Electric Company. Naturally the close interrelationship of companies is not exclusive to Hong Kong, although it has been a very notable feature of the colony's commercial structure. Moreover, it becomes something more than a passing interest when the linkages are ultimately traced back to the land market. The whole process of property development then is more than a vehicle for urban development, since it underlies the economy as a whole. Viewed overall, therefore, the Hong Kong economy in both the private and the public sector is highly conditioned by the operations of the property development sector. From the geographical point of view, the property process has promoted considerable urban change and development.

Until 1978, property development was proceeding relatively modestly, re-establishing itself after the collapse of the local stock market in the mid-1970s, and having to face the same problems of falling demand for property in the face of world recession, as those which blighted Brussels. Hong Kong Land in a joint development with a subsidiary of Jardine Mathieson, and the Hong Kong and Shanghai Banking Corporation, developed the World Trade Centre tower with office accommodation of 39,806 square metres (300,000 square feet), completed in 1975 (Estates Gazette, 11 October 1975). The development, on the foreshore of Causeway Bay and close to the Cross Harbour tunnel link to Kowloon, was one of many developed by Hong Kong Land, which began to acquire a large proportion of all the "prime" sites within the traditional central business district.

By 1978, the Mass Transit Railway Corporation was becoming a major developer and owner of property. For instance in a joint development with a local company, Cheung Kong, the MTR developed two office towers above their Admiralty Station in central Hong Kong, amounting to 105,513 square metres (794,000 square feet). In the following year, there was a rise in interest rates which might have been expected to put a brake on this development process, particularly as bank lending for property development was at a high level. In fact there was already a large amount of new offices in the process of construction and there were few signs that the cycle of growth was weakening. The shortage of new offices in 1979 was leading to an increase in rent levels and it was estimated that some 671,222 square metres (7 million square feet) of offices then planned would be on the market by the early 1980s. (Estates Times International Property Review, 1979). Much of the new development was outside the traditional CBD such as that on reclaimed land at Tsimshatsui East, which had a projected office total of 95,888 square metres (1 million square feet), with considerable additional development of hotel and retailing space.

A record number of new companies were registered in Hong Kong in 1980 with an additional 1,303 new foreign companies establishing offices in the colony. A growth rate of 12% in the economy in 1980 meant that, in contrast to much of the rest of the world, Hong Kong was a bouyant market for property development, with active sales of property and a rise in rental levels in the Central district of 65% between January and November 1980 (Estates Gazette, 21 February 1981). The extraordinarily active nature of the market was emphasised by the much publicised purchase of an office building for $HK998 million in January 1980 from Hong Kong Land by the Carrian Group, who arranged its sale for $HK1.68 billion some six months later, realising a profit of just over £60 million. The operations of the Carrian Group in the following three years were to reflect the changing fortunes of the Hong Kong property market as it began its decline. As an agent of change, however, it joined with other major property companies to alter dramatically the pattern of activity in the colony within a very short period of time.

Other areas were beginning to emerge strongly in 1981, including Wan Chai, Causeway Bay and especially Tsmishatsui East, with large amounts of office space coming onto the market, adding in total some 30% to the total office stock during that year. (Investors Chronicle, 3 April 1981). However, there were signs of a cyclical downturn in the market and a general sense of unease undermining somewhat the commercial confidence. Some of this stemmed from speculation concerning commercial rent control, as a possible extension of controls already extant in the residential sector, although such controls in the commercial sector and any subsequent deterrent of investment would hardly have been in the government's interest. It was also true that the colony could not remain immune from the effects of the more generally felt economic recession, so that 1979-1981 boom began to wane. One sign of the weakening of the market came in the declining value of the entitlement certificates (Letters B), given to leaseholders as an alternative to cash when land in the New Territories was taken back into government ownership. (Estates Gazette, 31 October 1981). The certificates gave prior entitlement to land released for development by the government and thus became marketable assets in their own right, acquiring values well above the alternative cash settlement offered by the government. At the peak of the property boom, the entitlement certificate could acquire a value some 2000% above the cash offer and were sought after by property companies to be added to their folios for future use. By the end of 1981, however, their market value was falling rapidly towards the level of cash settlement offered by the government.

Development was however continuing, largely because of the in-built momentum from the previous years. For instance, at eight MTR stations, there was the scheduled completion over the period 1982-1987, of some 652,044 square metres (6.8 million square feet) of commercial space and 575,333 square metres (6 million square feet) residential space (and the decline in value of residential property had been just as spectacular as that of the office sector). (Estates Gazette International Supplement, 1982).

Despite the warning signs, however, rent levels in Hong Kong were still exceptionally high. A Richard Ellis survey of world rent levels for prime 5,000 square feet offices showed that Hong Kong could still command the highest rent in the world - slightly in excess of £30 per square foot per annum, a level which was more than the £30 attained in New York and £27 in central London. (Financial Times, 12 February 1982). The reluctance to accept that the market was in decline was illustrated by continuing land dealings, notably that by Hong Kong Land in February 1982 involving what was officially termed Inland lot 8668, but more popularly known as Connaught II. The lease of the 1.3 ha. waterfront site was purchased for HK$4.78 billion (£436 million). To be developed by 1986, the development company planned 122,738 square metres (1.28 million square feet) of lettable office space with transport terminii, retailing and new premises for the Hong Kong combined stock exchange. Hong Kong Land's holdings in central Hong Kong already amounted to over 2½ million square feet (239,722 square metres) of property and the acquisition of this lot of land allowed the consolidation of other holdings to become a very large complex to be renamed Exchange Square. The stature of Hong Kong Land was not in doubt but in May 1982 the Financial Times commented "all eyes now appear to be focused on the property traders who appear most vulnerable and the property sector appears to be trading carefully in the hope that no-one trips up and deals a nasty blow to the confidences which, in Hong Kong permeates everyone and everything". (Financial Times, 21 May 1982). Continued urban development therefore was seen as conditional on confidence within the property market, re-emphasising the essentially mutual reinforcement of interests and values within the system.

There were, however, further difficulties to be faced. Certainly the growth of offices had been precipitous and resulted in an over-supply by 1982, as Table 4.3 indicates.

Table 4.3: Hong Kong Office Development 1978-1982

	Office Space Completed (sq.m.)	Vacant (year end)	% taken up
1978	184,400	131,000	158
1979	178,300	78,700	125
1980	296,700	202,900	42
1981	319,300	301,900	64
1982	485,000		

Source: Hong Kong and Shanghai Bank, quoted in Financial Times, 21 June 1982.

The percentage of offices taken up had been well below office completions, whilst the amount of new space actually becoming available by 1982 was continuing to increase. A 60% growth in office stock over that existing in December 1981, was planned for the end of 1984. Under normal conditions, the cycle of property development would have begun a downturn, but in Hong Kong other abnormal circumstances assisted it on its way. The political future of the colony, subject of much speculation, was discussed by the British prime minister with the Chinese government in Peking in September 1982. Although no decisions were reached, the effects of the meeting were felt in Hong Kong both on its Stock Market and in its property sector. Rents at the end of 1982 fell to 20-25% below their 1981 peak. In 1982, plans existed for 1.9 million square metres (20 million square feet) of new development by 1987, but by 1983, this was revised downwards to 1.25 square metres (13 million square feet). It was hardly surprising that property companies began to suffer just as they had in the UK in 1973-1974. It was not possible to stop all development, however, since heavy penalties were payable to the government if sites acquired for development were not developed within specific time periods. The traditional central zone of Hong Kong also maintained its value, but it contained in 1982 only one third of the office space, compared to the two-thirds in Central/Causeway Bay areas in 1979.

At this point it is worth pausing to consider the nature of those who had been responsible for the explosive growth of Hong Kong's office sector. Two companies, Hong Kong Land and Carrian Investments both warrant a closer examination since they were the promotors of much of the urban change. The precarious power of property interests is vividly illustrated by their activities, made all the more poignant by the effect that the market collapse of late 1982 had on them. In autumn 1982, several property companies had severe financial problems. Eda Investments, in which fifty-seven banks had financial interests went into temporary receivership and ultimate bankruptcy (Financial Times, 5 November 1982 and 1 February 1984). Carrian Investments also had liquidity problems brought about by the non-realisation of profits on certain property dealings. Its 'lifecycle' indicates the complexity of the land development process and the power of the property interests described in Chapter Two. The Company's origins were in Malaysia and Singapore, from where it emerged in 1980 to acquire a small property company, Maitton Enterprises, operating in Hong Kong. The spectacular profit made on a land deal in 1980 established the company as a major participant in the property market of the colony. Property developments included projects in North America, notably in Oakland, California where it was developing the tallest office tower on the west coast containing 305,688 square metres (2.3 million square feet) of lettable space. By May 1982 the company owned

230 ha. of land in various stages of development in Hong Kong, and in the twelve month period from late 1980 to late 1981, its net assets had increased from HK$1.19 billion to HK$5.5 billion. (Financial Times, 21 May 1982). As an agent of large scale urban change, therefore, the company was of very considerable importance.

It joined with Hong Kong Land to develop certain projects in the colony including the old Miramar Hotel in central Kowloon, for which the consortium, including the two companies, paid HK$2.8 billion to house a redevelopment scheme of 105,477 square metres (1.1 million square feet) of mixed office and retail development. In a review of the company in mid-1982, the Financial Times was able to conclude that "property trading looks set to contribute a substantial proportion of profits and Carrian is expected to maintain its tradition for the spectacular." (21 June 1982).

Such expectations were not to be disappointed since the company was to accompany the demise of the colony's property market which collapsed in a remarkably short period of time. An example is provided by the prices paid for separate floors of the Far East Finance Centre, on the edge of the Central District of Hong Kong. Average prices in February were HK$4,300 per square foot, but progressive falls through the year resulted in a September price of HK$2,550 per square foot, a fall of 59% in eight months. The complexity of Carrian's dealings makes a precise analysis of its losses difficult, but by October the company was in considerable financial distress (Financial Times, 5 November 1982).

Even the almost monopolistic Hong Kong Land was not immune from the problems. The dominance of the central Hong Kong market brought criticism such as, "the (Hong Kong) Land Company has the power to decide what location constitutes price, merely by deciding whether or not to be there. It can set rents by broadroom decisions rather than by taking account of the state of the lettings market." (quoted in Financial Times, 21 June 1982). Even if its indirect power over the urban land market is thus exaggerated, the company undoubtedly had considerable influence over the office development process by its direct activities. It was estimated in 1982 that of the 431,500 square metres (4.5 million square feet) of prime office floor space due to be completed from then until 1987, some 239,722 square metres (2.5 million square feet) (55.5%) belonged to Hong Kong Land. It would then hold approximately 479,444 square metres (5 million square feet) of prime office accommodation. The company had acquired links back to the trading heritage of the colony by an agreement made in 1980 with Jardine Mathieson, an original trading company which has become one of the colony's largest companies, to establish 40% cross-holdings in each other's operations. The picture thus presents itself of major companies shaping the urban form of the colony with complex inter-linkages binding them together.

The land market had reached such heights by 1981 that only a few very large companies were in a position to decide the type and form of urban development in the central business district, since they could out-bid other interests in the government disposals of development land.

Hong Kong Land survived the property market crash although not without difficulties, including the need to arrange a major re-financing operation in January 1983 following major losses in 1982-83. The selling of some of its disparate interests was a part of the reorganisation of the company carried out through Jardine Mathieson's interests in it. It is interesting to note that a direct descendent of the original Jardine family was destined to become chairman of both Jardine Mathieson and Hong Kong Land (Financial Times, 11 August 1983), thus re-asserting the power and influence of the original trading company in the colony.

In Hong Kong the development cycle illustrates the necessity to view the office development in its temporal context. The period of rapid development was remarkably short, until various pressures, not least political, meant that it did not conform to any normal expectation of a building cycle. Instead rapid acceleration of development over three years was followed by severe collapse. During that time, however, the office distribution of the colony was transformed, as indicated in Figure 4.2. Subsidiary districts have mushroomed in addition to the spectacular growth of the traditional centre. For those working in the offices, or in commerce and retailing, their life-style has been shaped and conditioned by the decisions of major companies motivated solely by company profits. Ambrose and Colenutt in their critique of the British property system suggest that "trading and speculating in property are some of the main activities of the finance capital which is, itself, near the heart of the capitalist system" and "we argue that the process of creating and maintaining high property values has side-effects that are inevitably and invariably damaging to the interests of non-property owners, especially to local low-income residents and workers." (Ambrose and Colenutt, 1975, p.134-5). Whilst the present study can offer no direct evidence to support the latter claim, the Hong Kong office development process was undoubtedly fuelled by property interests themselves creating their own land values. The government in its turn was so financially dependent on the process that it was not in a position to exercise the control necessary.

THE IMPACT OF THE PROCESS SUMMARISED

Both Brussels and Hong Kong stand out as examples of cities which have borne the brunt of the office development cycle. In neither case was there a continuing balance between demand and supply for office space and for varying reasons both suffered from vast over-supplies of office space. Speculative office building was the major form of development in both cities since the property industry had every reason to believe that there would be a demand for their new property. The property companies themselves, having created the artificial conditions of the land market have shown that they are particularly vulnerable to outside influences which disturb the market. Public control of the development process which might have been expected to be exerted in the interests of the community at large was in one case weak and prevaricating, and in the other contradictory to the financial interests of the government itself.

More generally the property development process has changed cities markedly. Urban areas which have gradually evolved over long periods of time have been transformed irrevocably and public control over that process has been often quite minimal. Short-term cycles have been especially traumatic to the urban environment since the urban system has to endure a period of adjustment to new development, affecting both the economic and social life of the city. The case study of Brussels illustrates this particularly well, as the property industry wrought irreparable damage to the city. The effect was that "Bruxelles est devenue une ville de tours qui éclipsent les monuments historiques". (Comhaire, 1975, p.25).

Chapter Five

OFFICE DEVELOPMENT PATTERNS AND PROBLEMS IN THE UK

The pattern of office development in Britain is a result of an interplay between the demands of office users, the requirements of office investors and the effects of the office control system as it has operated since 1945. The resulting domination of the distribution by London is not surprising since its place as a prime international centre of office activity above all for finance, has been increasingly reinforced during this period. Despite the cyclical pattern of demand induced by economic boom followed by recession, therefore, London's position has always been strong, providing the secure and reliable investment base so essential to the financing institutions, discussed in Chapter Two. Nonetheless, despite the pre-eminence of London, the office market in the UK is still a much more dispersed one than those of many other European countries and several provincial cities have seen considerable activity in terms of new office construction. The reasons for this include the general tendency for a degree of decentralisation of more routine activities, for instance in the banking and insurance sectors, towards office centres outside the capital. An analysis of the effects of these trends indicates, however, that certain areas, notably the outer Metropolitan area of the south-east and the western corridor stretching outwards from London to Bristol, have particularly benefited from the decentralisation process. These trends are discussed in the first part of this chapter before returning to a detailed appraisal of the situation in London, which has seen three distinct office cycles since 1945 (see Barras 1984), culminating in a clear questioning of the hitherto accepted process of development, as illustrated by the proposed large-scale development of offices bordering the South Bank of the Thames.

GEOGRAPHICAL PATTERN OF THE BRITISH OFFICE MARKET

The geographical distribution of the office market in England as reflected in Table 5.1 is clearly very unbalanced with over half

DOI: 10.4324/9781003174622-5

of the commercial office space being in the South East region. Of the 24.8 million square metres of offices in the South East, it is Greater London which, not surprisingly had the largest share with almost 69% of the region's total and 37.6% of the total of the country as a whole. If attention is focused on larger office units, i.e. those whose floor area exceeds 1000 square metres, then the importance of Greater London becomes even more evident, since it houses 44.5% of all such offices in the country. Of the English planning regions, moreover, it is apparent that outside the South East, it is the North West, whose office market naturally centres on Manchester and Liverpool, which is of some relative importance. Indeed it is the only region to have more commercial office space than the outer metropolitan region of London. At the other end of the scale, it is worth noting that East Anglia, admittedly dominated by rural interests, has the least commercial office space (1.3 million square metres), whilst the Northern region, of which Newcastle is the centre, has only 1.8 million square metres, with the smallest average size of office in the country.

Table 5.1: Distribution of Commercial Office Space in England, 1982

Region	Total (million sq. metres)	% total	Office units over 1000 sq. metres	% Nat. Total	Mean size of office unit (sq.m.)
Northern	1.8	3.96	0.7	3.18	195
Yorkshire and Humberside	3.2	7.05	1.2	5.45	198
North West	5.5	12.11	2.5	11.36	220
East Midlands	2.1	4.62	0.8	3.64	207
West Midlands	3.5	7.71	1.4	6.36	216
East Anglia	1.3	2.86	0.6	2.72	217
South East	24.8	54.63	13.4	60.91	303*
Greater London	17.1	37.67	9.8	44.55	349*
Outer Metropolitan	4.5	9.91	2.2	10.00	259*
Outer South East	3.2	7.05	1.4	6.36	208
South West	3.2	7.05	1.3	5.91	207
England Total	45.4	99.9	22.0	.33	253

Source: Commercial and Industrial Floorspace Statistics 1979-82. (HMSO 1982).

Table 5.2: Office Floor Space increase in English Planning Regions, 1979-82

Region	Increase 1979-82 (million sq. metres)	% increase 1979-82	% total increase
North	0.1	5.88	2.63
Yorkshire and Humberside	0.3	10.34	7.89
North West	0.3	5.77	7.89
East Midlands	0.2	10.0	5.26
West Midlands	0.3	9.38	7.89
East Anglia	0.1	8.33	2.63
South East	2.2	9.73	57.89
of which:			
Greater London	1.3	8.23	34.21
Outer Metropolitan Area	0.6	15.38	15.79
Outer South East	0.3	10.34	7.89
South West	0.3	10.34	7.89
TOTAL	3.8	9.38	99.97

Source: Commercial and Industrial Floorspace Statistics 1979-82. (HMSO 1982).

Table 5.2 emphasises very clearly that the general regional pattern is by no means likely to change. Instead the recent trends have been to boost the amount of commercial office space in the South East by nearly 10%, including a 15% increase in the period 1979-82 in the outer metropolitan area - with over half of the total of new office space being built in the south east. This general tendency for the South East to strengthen its position comes as no great surprise when one considers two factors. One is the fact that restrictions, in the form of office development permits were lifted at the beginning of the period in question and it could be argued that a pent-up demand existed awaiting new developments. A more convincing explanation however, was in the attitude of the property market interests themselves, since the strongest demand with the most secure return is perceived as being in the south-east rather than elsewhere, and the office development process operates in response to such opinions.

The regional distribution of major funds, including both property unit trusts and life assurance funds emphasises the spatial bias of such investment, particularly when freed of regional controls. Table 5.3 shows the total investment pattern of each major investor, together with the distribution of their property assets between the various sectors, including offices.

Table 5.3: Percentage Distribution of Property Assets of Major Property unit Trusts and Managed Funds by Sector and Location

	Sector (%)				Location (%)						
Property Unit Trusts	Office	Shop	Ind.	Agric.	London	Rest of SE	Wales/ SW	Midlands	North	Scotland	Over- seas
Fleming	38	31	25	6	43	26	10	10	7	4	8
Hanover	45	13	42	-	67	16	-	6	11	-	-
Hill Samuel	49	32	19	-	39	28	7	4	19	3	-
Lazard	50	29	21	-	37	38	5	5	5	10	-
New Court	48	23	22	7	20	50	-	18	12	-	-
PFPUT	46	33	-	21	60		4	10	7	9	10
PUTPAGS	31	25	19	25	24	70	6	-	-	-	-
Schroder	42	36	20	2	19	39	10	13	16	3	-
Managed Funds											
Legal and General	49	39	9	3	44	15	14	6	12	3	6
Norwich Union	78	20	2	-	32	23	7	26	9	3	-
Standard Life	47	25	28	-	34	22	14	10	10	10	-
Prudential	40	42	12	6	41	24		29		6	-

Source: Investors Chronicle, 6 April 1984

An analysis of office rents as presented on the annual reports of major property companies indicates the primary centres of demand for offices in the UK. Whilst rents are admittedly also a reflection of supply since oversupply depresses rent levels and shortage of space increases rents achieved, in broad terms, the rents asked and paid in a centre give an indication of its position relative to other centres. It is of course difficult to present precise rental levels for any one city since they may vary widely even within one city and indeed dependent on the activities of agent, leasor and leasee, they may vary within one building. Nonetheless, despite these qualifying comments, a hierarchy of office centres can be identified according to prevailing rent levels.

Recent rent levels for prime office space in the City of London have reached £2.80 per square metre (£30 per square foot), which are amongst the highest in the world. Whilst these are very high indeed, other London locations, particularly the West, such as Hammersmith, have reached £1.20 or more per square metre, only slightly above other outer London centres such as Croydon and Bromley. The westward extension of office demand is particularly emphasised by the position of Reading in the rent "league", since rent levels for new office space in the town have reached those of West London prime office accommodation. Outside the western corridors, the highest rents are achieved in the major provincial centres such as Birmingham, Manchester, Bristol, Cardiff and Leeds, whilst in Scotland the cities at this level are Glasgow, Aberdeen and to a lesser extent, Edinburgh. Added to these are major centres in the outer South East such as Brighton and Bournemouth. In all these cases, however, rent levels are rarely more than 50% of those achieved in West London and may be as little as 20% of the levels of the City of London. The second tier of provincial centres with somewhat lower rents includes smaller provincial cities such as Nottingham, Sheffield and Coventry with rents 10-12% of those in the City.

Inevitably such analyses are generalised and short term since over-or under-supply may distort the levels for a particular period of time. One area where this is an important factor is the East Midlands, where the development cycle of the 1970s left an over-supply which has tended to reduce rents quite considerably. The worst affected place is Leicester where in 1983, a supply of vacant space of 55,000 square metres was reported with long term vacancy of some office blocks and rents as low as £0.09 per square metre (£1 per square foot). (Financial Times, 20 May 1983). It is inevitable that such areas will remain unattractive to institutional investment at least until the next upturn in the economic cycle ensures a take-up of vacant offices, an increase in rent levels and a subsequent upturn in the office development cycle. Despite such areas, however, this broad generalisation remains of a hierarchy, incorporating in descending order, a particularly high rent City market, a secondary London and outer metropolitan market, with an emphasis on the west of the capital, a first level provincial market encompassing most large provincial centres and a secondary provincial market with smaller regional centres.

The yields on investment in the high rent areas are of course likely to be low, simply because of the cost of investing in such areas, but they offer the long-term security with relative certainty of increasing rent levels which the secondary centres, with higher initial yields on investments may not necessarily offer. The result is that the hierarchy as outlined above tends to reflect the investment activity of major institutional investors, which has already been summarised in Table 5.3.

The increasing attraction of outer Metropolitan centres is principally explained by a tendency to decentralise from the capital, particularly involving the movement of the more routine office functions to lower cost space. This trend is not of course, unique to London. Catalano and Barras (1980), in their detailed analysis of the office development process in Manchester indicate the early start to the loss of total employment by the centre (Manchester SMLA), dating from the 1950s and accelerating thereafter. Within this pattern their figures demonstrate the absolute decline of total office employment, particularly during the late 1960s. (See Table 5.4).

Table 5.4: Office Employment in Manchester and the North West Region, 1961-71

	1966 (000s)	1971	% Change Total Office	Clerical	Admin. and Managerial
England & Wales	4986.9	5439.9	+9.1	+4.4*	+18.9
North West Region	650.2	691.5	+6.3	+2.8*	+18.8
Manchester C.B.	121.9	107.6	-11.7	-16.9	+4.2
Manchester City Centre	67.4	59.2	-12.1	-18.0	+10.6

(Source: From Catalano and Barras, 1980, Table 6, p.9).

*Includes some other office occupations.

Whilst Central Manchester itself was losing office employment, outer suburban centre such as Bury, Bolton, Stockport and Wilmslow saw increases between 16.6% and 25.5% in office employment during the same period (Catalano and Barras, 1980, p.9). The other important trend was the increasing concentration of decision-making in the centre of the conurbation, illustrated by the growth of administrative and managerial jobs within the core, although that did not preclude a simultaneous suburbanisation of some of these office functions. The trends in Manchester are very important for two reasons. Firstly they reflect, albeit on a small-scale, the patterns of demand for office space in the South East, where suburbanisation is joined by a more widespread dispersion of office activity, and particularly of routine clerical office functions. Secondly, the concentration of managerial positions in the core, accounting for an increasingly larger proportion of total office jobs has the effect of maintaining the demand for office space, despite the fall in office employment, simply because the space demands per worker at this level in the office hierarchy are greater than those for clerical workers. In the context of the London office development market, it is this which goes some way to explain the maintenance of demand for new office space by office functions at the core, although as will be discussed below other factors, such as the growth of foreign banking activities, have also been of considerable importance.

OFFICE RELOCATION

The extent of recent relocation of offices from London is difficult to establish, particularly since the cessation of publication of statistics by the Location of Offices Bureau. Even these, available until 1979, included only a part of the moves undertaken. However, a survey by Jones Lang Wootton in 1983, examined major moves from London in the period 1979-82, involving relocations of more than 100 jobs. (Jones Lang Wootton 1983). The survey indicated 62 relocations during this period, involving 54 private companies, and eight public sector organisations, to which should be added a further seven relocations by central government departments, not included in the survey. 13,000 jobs were thus relocated, releasing 204,000 square metres (2.2 million square feet) of office space in central London. The recipient centres of these relocations were much the same as those during the existence of LOB, namely the outer metropolitan area and the rest of the South East, with the region as a whole taking two thirds of the relocations.

Notable amongst the functions moving from London was the insurance industry with nine relocations. This is a continuation of the trends in the 1970s which amongst others, took Eagle Star to Cheltenham and Zurich Insurance to Portsmouth whilst a future move will take part of Commercial Union's City activities to Basildon in 1985. Perhaps of greater significance, however, is the fact that manufacturing industry was also prominent in the relocation of offices from London during this period, with chemical and related industries and the engineering and electrical sector both recording eight moves and printing and publishing, seven. The movement of office activities related to manufacturing accounted for 52% of the move in this survey, a figure very much higher than the 34% of the relocations recorded by LOB from 1964-79. A final important point emerged from the survey and that was the fact that there was a tendency for relocation to be somewhat further afield than previously - 76% of moves 1964-79 according to LOB were to the South-East, as opposed to the 66% indicated here. The inference to be drawn is that the office activities related to the manufacturing sector are increasingly moving back towards existing centres of production outside the South East. (Financial Times, 14 October 1983). International Harvester's move of its UK headquarters to Doncaster from London and Tube Investment's transfer of many of its head office functions to Birmingham, the focus of its West Midlands production activities are two examples of such moves.

THE IMPACT OF RELOCATION ON THE OFFICE MARKET

The relocation of office activity has often served to stimulate a local office market. The relocation of the 1960s and 1970s occasionally led to an over-stimulus of the market, providing a surplus of office space in response to an anticipated continuation in demand which in fact did not materialise. One centre so affected is Bristol which was an important location for financial institutions, both banking and insurance, moving from London in the 1970s. The popularity of the city with developers, sustained by the confidence stemming from the arrival of such prime office functions, led to a surplus of office space of the order of 100,000 square metres by the mid- to late-1970s. By 1983 the situation had improved somewhat, although there was still 58,000 square metres available and approximately 50,000 square metres under construction. (Financial Times, 25 March 1983). Development of offices in Bristol, however, is not confined to its traditional commercial areas since the Aztec West High Technology Park, developed by the electricity industry's pension fund, Electricity Supply Nominees, also incorporates speculative office development, with the commencement of a 3,716 square metres (40,000 square feet) speculative development in 1984. Aztec West's location on the northern outskirts of the city, close to the M4 and M5, with Bristol Parkway station in the same vicinity, shares the general advantages of Bristol as an area for relocating offices, particularly with the regular high speed inter-city 125 service linking it with London. It is equally true that it lies within, although at the extremity of the favoured western corridor outwards from London.

As a preferred sector for relocation this particular corridor has had considerable success and the reasons for its popularity are not difficult to discern. Whilst office rents in the corridor themselves may still be high by provincial standards, and are still rising because of the sustained demand, the communications advantages, coupled with the possibilities of employees living in pleasanter surroundings than Greater London, are very powerful. Successive surveys during the 1970s demonstrated the social and economic benefits to be had from such relocations, particularly to centres in the outer south-east (see for example, Bateman and Burtenshaw 1979).

The western corridor has the added impetus of having attracted much high technology industry such as the computer companies, Digital and Hewlett Packard to Reading. Much of this style of activity, often involving micro-technology requires specialised accommodation. There is no doubt that this can easily be provided in campus style developments, but these are already giving rise to planning problems in the face of increasing pressure on available land. Nonetheless, the progressive image of the zone is an important one, particularly for large, often international companies, seeking prestige locations outside London. Recent surveys suggest that the demand for office space in the corridor stretching from Hammersmith to the county of Avon is of the order of one million square metres (report of Knight, Frank and Rutley, reported in The Times, 24 January 1983).

Reading has benefited particularly from the increased demand for office space in the western corridor, as reflected in current rent levels for new offices which equal those of West London. Examples of relocation include the move of British Telecom's Yellow Pages division from North London to a new building providing 6500 square metres of office space, financed jointly by the London and Edinburgh Trust and the Civil Aviation Pension Fund. (Estates Gazette, 4 February 1984). The demand for office development is such that the local authority has refused permission for several schemes in the recent past on the grounds that development of one site would prejudice larger scale development involving adjacent sites. A very large scheme is under consideration, to be developed jointly by Prudential Assurance and British Rail on the Reading station site. In February 1982, the local authority refused permission for the development of 34,350 square metres (370,000 square feet) of new space, of which 78% was to be speculative development, with British Rail and the Post Office occupying the remainder equally. The district plan for the area has been modified to incorporate 23,225 square metres (250,000 square feet) of offices on the site, although this is still less than that proposed by the developers. The general tendency therefore is for there to be pressure from the development interests on the town, indicating its important place in the general hierarchy of investment.

It is worth recording that there can be considerable gain for a local authority in a situation such as that of Reading where demand for development is strong. It is particularly well illustrated by the details of a 1.58 ha (3.9 acres) site at

King's Road, Reading (reported in detail in Estates Gazette, 15 October 1983). A property development group, MEPC, purchased the site from the major local authority involved, Berkshire County Council, for £9.3 million. The scheme granted planning permission involves 32,500 square metres of offices, plus both new residential development and limited residential development associated with the restoration of listed buildings on a part of the site. The minor local authority, Reading Borough Council have entered into agreements involving their paying MEPC some £2.0 million to provide certain highway facilities and some residential development. The whole scheme will be financed largely by the Legal and General insurance company. It is worth recording that an earlier scheme involving the purchase of the same site by MEPC for £12 million in 1981 and the use of the site almost entirely for offices, did not proceed because of the objections by Reading Borough Council which was attempting to restrict office development, in contradiction to the county's wish to proceed with the development of the site.

The case of Reading illustrates well the pressures both from the institutional investors and from office users. The unequal distribution of demand from both sources on a national scale is no less demonstrable and the lack of national controls on the development process is a serious deficiency at a time when there is a clear demand for new office development. The difficulty of control and the conflicts of policy are particularly well illustrated by London itself, however, and the next part of this chapter will examine the patterns of development in the capital, concentrating especially on the areas under pressure on the South Bank.

LONDON - THE GENERAL SITUATION

The general dominance of London in the UK office market has already been emphasised, but Barras provides further evidence of this strong position, (Barras 1984). Whilst 30% of the total national office employment is in London, he points out that 40% of the commercial office employment is in the capital. By value, 80% of national office rateable values are in London, 60% in the Cities of London and Westminster alone (Barras, 1984, p.35). The demand for office space is still high, despite the decentralisation trends discussed above. There are a number of reasons for this, including the increasing national concentration of headquarters offices, despite relocation trends, and the increase in office space per worker, an extension of a trend noted by Daniels of an average increase in the GLC of 1.2% per annum from 1961-76 (Daniels 1975). A further very important factor is the phenomenal increase in foreign banking activity in the City and subsequent demand for office floor space. A series of surveys since 1962 have analysed this growth in detail (see for instance Noel Alexander Associates, 1983). In 1982 alone, 31 new banks were established in London, whilst two left, a net increase of 29. The net pattern of foreign banking is shown in Table 5.5, in which the presence of foreign banks is shown at five yearly intervals since 1960 to 1980, together with the latest available figures.

Table 5.5: Growth and Origin of Foreign Banking Interests in London, 1960-82

Year of Origin	USA	Europe	Japan	Arab	Others	Total
1960	8	27	8	1	29	73
1965	15	32	11	1	40	99
1970	37	51	12	3	55	158
1975	58	90	23	7	79	257
1980	71	141	24	19	128	383
1982	77	153	29	26	143	428

(Source: from Noel Alexander Associates (1983). New Foreign Bank Offices in London - 1982).

It will be seen that the banking 'invasion' of London during this period involves phases of establishment of banking interest from various parts of the world. The European banks were relatively well established by 1960, although their number has still grown enormously since then, whilst the American banks have become gradually established since the 1960s. The Japanese also became particularly well represented in the 1970s whilst a more recent influx has been from the Arab countries of the Middle East. The physical proximity of the banks one to another is particularly important and accounts both for their presence in the City and a degree of clustering within the City itself.

A further stimulus to demand for new office space arises from the need to replace obsolete premises with new buildings into which modern communication facilities can be more easily accommodated. Whilst refurbishment of premises may partially satisfy this requirement, this may be prohibitively expensive (see Chapter Eight for a detailed discussion of this topic). Modern communications are a *sine qua non* of money market operations, underlining the particular needs of these office functions for modern office space.

The demand for office space exists therefore in London, but a second demand also exists from the development industry itself. Whilst yields may be comparatively low, as indicated above the security of investment in London in general and the City in particular, is considerable and provides just the right type of outlet for the funds of the more cautious investing institutions. This demand coupled with the demand from office users combines to ensure that the London office market is still the most important in the country.

As Barras points out, London has undergone three development cycles. (Barras 1984, p.45). The first cycle involved the development of war-damaged sites in the City during the 1950s and early 1960s. Once the supply of these sites had been exhausted by the early 1970s, attention turned during the second cycle to the redevelopment of obsolete offices, often leading to considerable net increases in floor space. Barras terms this, the 'intensification' of the central area office stock, reflecting increased competition for the high profit development sites. The third cycle dates from the late 1970s and has tended to focus both developers' and planners' attention on sites on the fringe of the City, in inner city areas not traditionally seen as office quarters. The South Bank developments and proposals are seen as a natural extension of this cycle which in general involves very large scale projects, since the sites themselves are very much larger than those previously available in the City.

It is the very large scale of these proposed developments which makes them so important to an analysis of the London office market. They do not represent a replacement of existing offices since they frequently involve areas in which manufacturing or dockland functions were predominant. At the same time, they represent a potential large scale transformation of the areas concerned and it is this factor which accounts for the political and community interest in many of these proposed developments.

LONDON'S SOUTH BANK

London's South Bank has been the scene of a series of confrontations within which the inevitability of large scale office development, propelled by the property industry, has been called into question. The three largest schemes involve sites at Hay's Wharf, between London and Tower Bridge, at Coin Street, to the east of Waterloo Station and adjoining the National Theatre and further up river the on the south side of Vauxhall Bridge. The last development involves more than 140,000 square metres (1.5 million square feet) of offices in a series of phased developments, financed by Kuwaiti interests. The development at Hay's Wharf involves 68,750 square metres of offices with 7,800 square metres of other accommodation, both retail and residential. Once again, it is overseas interests, in the form of the St. Martin's Property Corporation, owned by the Kuwaiti government, which are concerned with the Hay's Wharf site. This development has the advantage of close physical proximity with the City, being across the river from it and could act as a natural overspill for the more traditional office areas north of the river. Nonetheless, its development represents a very considerable change of function for the area concerned, and it is this change of function which is central to the conflicts over the most controversial of the South Bank sites, that at Coin Street, which will now be considered in some detail.

Coin Street - an example of Community Control?

The proposed development of offices at Coin Street has proved to be extremely contentious and has involved a series of both public enquiries and judicial actions, all of which underline both the scale of the development proposed and the fears of its potential impact on the community. The situation is exacerbated by political considerations involving a change of political control of the Greater London Council (GLC) during the period of negotiations concerning the site, the need to consider local political views as expressed in local plans, and finally a conflict between central government and the GLC. Each of these elements will be considered in turn, starting with a consideration of GLC planning policies.

The Greater London Development Plan, as finally approved in 1976 sought to encourage offices in so-called preferred locations, which were seen as those with good public transport accessibility. In addition, it contained a policy to seek where possible, planning gains for the community as a condition of office development and also to take account of labour availability. These policies were adopted in the knowledge that employment within the GLC was declining and in the desire to ensure that such a trend was controlled. Within central London, the preferred locations included the City and most of the main line rail termini, together with other locations seen as having good transport accessibility. Certain locations on the northern and eastern fringes of the City were included as was the South Bank area of the two local authorities of Lambeth and Southwark, although the GLDP made only indicative recommendations for office development in all such areas, rather than precise recommendations as to site.

The local plans of the London Borough Councils were intended to provide amplification of the GLDP, but even as early as 1977 when Lambeth's Waterloo District Plan was approved, it was evident that local plans might well contain policies opposed to those in the GLDP. When in 1981, however, there was a change in political control of the GLC, a more restricted policy was adopted towards offices, which was much more in line with the limiting policies adopted in local plans by the borough councils. Indeed the new policy of the GLC specifically supported the provision for offices as laid out in adopted local plans and rejected the notion of planning gain as a justification for approval of office development outside those areas designated in local plans. In general, office development throughout central London was to be considered acceptable only if it involved the modernisation and redevelopment of existing office premises, if offices were ancillary to and necessary to the proper functioning of industry and commerce, if they contributed to the improvement of public transport facilities and exceptionally if they were for a named occupier seen as essential to supporting London's economy.

It could be argued that local plans were even more restrictive in their attitude towards office development. The policy of Southwark, in which part of the Coin Street site and all of the Hay's Wharf site is located, is a case in point. In its office policy review in 1982, policies were adopted which were entirely prohibitive towards office development on the South Bank and the area of London Docklands Development Corporation (largely downriver from Southwark). Elsewhere in North Southwark, policies in line with those adopted by the GLC in their review were affirmed. The local authority was willing to allow office development which was of benefit to the local community, but this would not include that in the north of the borough on the South Bank, since not only would such development undermine the local community and be in conflict with it, but it would be doing little to assist in the employment needs of the local population. This is confirmed by the local authority's own survey in 1979 of place of residence of employees in new offices constructed since 1970 in the borough, since only 9% of employees working in such offices were local residents. The restrictive policies were adopted in the light of an office pattern in the borough which showed considerable imbalance with, in 1981, 78% of the borough's total of office floor space of 980,728 square metres, being in its north west planning district, adjacent to the South Bank. Of this total, however, only 2.9% was seen as being "local service" floor space.

In summary, therefore, the Borough adopted a very strong policy against offices, rejecting the notion that the north of the borough was a natural extension of the City of London and preferring instead to cater for the needs of the local community in terms of low rise housing and the local economy and environment in terms of protecting and enhancing surviving industrial areas.

Prior to the change of policy by the GLC, however, a property company, Greycoats London Estates, had acquired an interest in the 5.26 ha.site on the South Bank, with a promise that the GLC would in time sell its own site interests, originally acquired in 1953, to the property company.

The site was the topic of a six-month public enquiry in 1979 after which the Secretary of State for the Environment refused permission for either the office proposals put forward by the property company concerned or the alternative housing plans proposed by Lambeth Council and an amalgam of local interests called the Association of Waterloo Groups (AWG). In refusing permission, it was indicated that a mixture of uses on the site would be more favourably received. A new enquiry began in 1981, and was adjourned for three months following bitter disputes after the signing of the GLC's agreement to sell their interest in the land, immediately prior to the elections to the GLC which were to change its political control. The enquiry was eventually concluded in March 1982. At that time, the GLC and both borough councils, Lambeth and Southwark, supported the plan by the AWG for a mixed scheme of development as opposed to the Greycoats' (now Greycoats Commercial Properties) scheme which still involved 82,170 square metres (884,500 square feet) of offices.

The curious situation then arose of both the AWG and the Greycoats scheme being given planning permission, a situation challenged unsuccessfully in the High Court by the GLC in 1983. Finally in April 1984, Greycoat Commercial Estates abandoned its plans for the Coin Street site and sold its land interests to the GLC for £2.7 million. (Financial Times, 4 April 1984). The AWG scheme involves the construction of 400 dwellings, a park and some light industrial floor space, although the means by which the scheme will be financed have not yet been finalised.

The long saga of the Coin Street site poses a number of questions concerning the office development process. It is certainly true that the time involved in reaching a decision on the site has been very harmful to the local environment, inflicting a planning 'blight' on much of the area. It also calls into question the procedures by which complex planning situations such as this are decided. Certainly two extensive public enquiries did little to solve the initial dilemma and ultimately, by resulting in the granting of two planning permissions, resolved nothing. On the other hand, the local community did demonstrate that it was able to resist the demands of the office development industry in that it succeeded in causing the property developer to abandon plans for the area. Nonetheless the potential remains for conflict between the local authority in areas such as the South Bank and higher authorities, including central government. The final chapter of this book provides an opportunity to pursue this issue further.

AN ASSESSMENT OF CURRENT TRENDS IN THE BRITISH OFFICE MARKET

Even a brief review of British office property development reveals a pattern which is dynamic and subject to periodic shifts reflecting local over and under-supply. Nonetheless, despite the local conditions, it is possible to identify some important trends which are affecting the office development process. We have already looked at the process of deconcentration from central London. It seems likely that such trends will continue, reinforced by the willingness of institutional investors to fund developments in particular areas. The movement of foreign banks away from the city is a good indication that such a trend is important and has a high level of acceptance. The US banks have in this case shown the way. Three of them exemplify the possible areas to which relocation may take place, with the First National Bank of Chicago moving to Covent Garden, within London's West End, the Bank of America moving to Bromley (Kent) and the Chemical Bank taking its activities to Cardiff. Nonetheless such an exodus, although well established and involving prestige companies, seems unlikely to undermine the importance of the City itself. Prestige addresses in the City for banking, insurance and other activities are still a powerful attraction, and rent levels in excess of £2.79 per square metre (£30 per square foot) strongly testify to this. Nonetheless, other locations do seem likely to increasingly attract the attention of the institutional investors.

Within the capital but outside the City of London, for instance, there are important office locations which are proving popular with both occupier and investor. The areas to the west of the city provide obvious examples, with lower rents and increasingly prestigious addresses in their own right. The joint development by MEPC and Legal and General at 90, Long Acre in London's Covent Garden has attracted major tenants such as Ultramar and Sun Oil besides the first National Bank of Chicago, noted above. Rent levels for this building in the range (£16.50-£18.50 per square foot) are quite high, but they are significant savings on city rents. (Investors Chronicle, 28 October 1983). Other developments within London with such status include the 6,503 square metre (70,000 square foot) scheme developed by Townsend Thoresen opposite Victoria Station and the adjacent 1,858 square metre (20,000 square foot) office development by Greycoats over the station itself. It could be argued that such prestige developments are being made available in time to benefit from the next upturn in the economic cycle.

In the outer metropolitan area there is no doubt that the M25 motorway around London has already proved to be influential in locating new office development. In Kent, for instance, centres such as Sevenoaks, Orpington, Tunbridge Wells and Tunbridge all afford excellent links to both Heathrow and Gatwick *via* the M25 and such accessibility was no doubt important to the Bank of America's decision to move in 1983 to its new 14,493 square metres (156,000 square feet) of accommodation in Bromley some ten kilometres to the north of the M25, yet still able to enjoy its ease of access to the major international airports.

Further to the west, there are indications that the popularity of the western corridor is likely both to increase and to expand in terms of geographical area. Already the south of London considered above may be seen as merely an extension of the high accessibility corridor to the west. In addition, north Hampshire is sharing in the development momentum of the M4 corridor. Basingstoke for instance, has its major Eastrop office centre and a planned 9,290 square metres (100,000 square feet) headquarters for Sun Life Assurance. The highest rents are still to be found within the corridor itself, however, with offices in the Royal Borough of Windsor and Maidenhead achieving rents of over £1.39 per square metre (£15 per square foot), encouraged by an annual restriction of new office provision to 2,400 square metres. (Investors Chronicle, 23 October 1983). Elsewhere in the outer metropolitan area, Milton Keynes is developing as a major office centre, with 27,870 square metres (300,000 square feet) let in one year, going some way towards taking up a surplus of 37,160 square metres (400,000 square feet) which had been available in 1982. In general it seems fair to predict that demand in the South East, given the lack of government control, is sufficiently strong to continue to attract institutional investment on a large scale. Office development in the region generally offers the right level of security of investment demanded by the traditionally more cautious institutional investors. Additionally, once some of this type of investment demonstrated its confidence in the area, then others have followed in the mutually supporting process discussed in Chapter Two.

Elsewhere in Britain, two trends are of importance. One is a degree of decentralisation, or suburbanisation of offices, although not on the scale of that seen in London. The second is that of refurbishment of older premises, a trend which may also be identified in the capital. The two trends come together in the case of Edgbaston, in Birmingham, when refurbishment of offices dating from the 1960s has been a recent trend. The potential for major decentralisation to suburban locations may be seen clearly in both Birmingham and Manchester, although other cities such as Leeds and Liverpool have also experienced it on a more limited scale. The refurbishment of central area offices dating from the last century has been important in many cities, often encouraged by planning policies aimed at conserving the Victorian heritage of major industrial cities, such as that of the centre of Newcastle-upon-Tyne, where Grainger street and Grey Street are parts of a central conservation area, containing major commercial premises, including offices. In Liverpool, the building at Water Street/Castle Street, opposite the City's Town Hall has recently been refurbished to provide ground and mezzanine floor accommodation for a building society, whilst its upper floors have been let at £0.49 per square metre (£5.25 per square foot), which is a relatively high rent for the city. (Investors Chronicle, 23 October 1983). The attraction of refurbishment schemes is that they offer the opportunity to provide small to medium size office accommodation for the local or regional service sector, including professional services, banking and insurance. Indeed in the provincial sector generally, two markets may be identified, one being formed by local service demands for premises up to 1,500 square metres, whilst the second is for more substantial tenants with no local attachment, possibly moving from central London. Such a demand is more likely to be met by major new development of over 2,500 square metres rather than by refurbishment.

Table 5.6 summarises some trends in office construction in terms of those centres with the most rapid growth of office construction in the period immediately before and after the abolition of office controls. It will be seen that the western corridor is particularly well represented together with other centres in the South East. Suburbanisation outside the South East is represented by Trafford to the South of Manchester. It is also evident that the most rapid growth of offices also took place in small to medium sized centres, although one should not overlook the fact that in absolute terms some larger centres also saw some considerable growth. For example, the City of London saw a 143,600 square metre increase (4.1%) in the same period and Leeds, an additional 122,900 square metres, (19.4%). Nonetheless the table indicates the type of centres which demonstrated a strong potential for growth. The implications of such trends will be further considered in the final chapter of this book.

Table 5.6: Local Authorities with Office Floor Space over 150,000 square metres in which Growth in 1977-81 exceeded 25%

	Floor Space Estimate		% Increase
	1977 (,000s, sq.m)	1981	
Thamesdown (includes Swindon)	137.4	195.3	42.1
Coventry	218.1	268.1	38.9
Windsor & Maidenhead	131.0	173.0	32.1
Luton	131.0	171.2	30.7
Reading	153.2	199.1	29.9
Milton Keynes	121.9	156.2	28.1
Derby	178.4	227.4	27.5
Harrogate	128.0	162.0	26.6
Trafford	191.8	242.2	26.3
Brighton	155.8	196.2	25.9

Source: Department of the Environment, Floor Space Statistics, England and Wales 1974/77 and 1978/81, HMSO. (1978 and 1982).

Chapter Six

FRANCE - A CONTROLLED MARKET

The office property market in Europe has been particularly active in Germany, Netherlands and France, as well as in the somewhat extreme case of Brussels discussed earlier in Chapter Four. Both local and international investment have been important sources of finance in these European markets, promoting an extensive amount of office development particularly in established office centres such as Frankfurt, Dusseldorf, Amsterdam and Rotterdam. For a number of reasons, the country whose office development process warrants close scrutiny, however, is France. One of the principal reasons is that it is the office market outside the UK., which has been most subject to control by government. The effects of such control measures, operative now for nearly two decades, have been far-reaching and indeed could be said to have fundamentally re-shaped part of the office market and thus the distribution of offices as a whole. It will become evident, however, that government control has been directed at influencing the major regional market of Paris itself, since there is only a poorly developed office market in France. Direct planning legislation has had its effect on office development, but equally important have been the indirect effects of government policies and statutes, particularly since the 1981 presidential election and the consequent implementation of the policies of the new socialist government. New forms of taxation and the uneasy economic climate stemming from that electoral victory have generally depressed an office market, although some policies of the Mitterand Government, such as those favouring decentralisation from Paris, might reasonably be expected to have produced an invigorating effect on the office development process in the provinces. Government influence on office location however, did not begin in 1981 since a strong government policy affecting office employment both nationally and especially in the Paris region has been in operation since the mid-1960s and a full appraisal of the office development process will enable us to assess the success or otherwise of these policies.

DOI: 10.4324/9781003174622-6

SPECIAL CONDITIONS OF THE FRENCH OFFICE MARKET

The peculiarities of the French tenure system require a brief examination in order to assess the trends in the market. In a general sense, speculative office building (*bureaux en blanc*) is a new phenomenon in France where owner-occupation of office property has been traditionally far more common. Only since the arrival of British property interests in the late 1960s and particularly the early 1970s, has the notion of holding property as a long-term investment gained a strong foothold. Since 1938, a system of co-ownership (*co-propriété*) of property has existed, originating in the joint ownership of residential property as a response to an acute shortage of such property and first given a legal standing in Grenoble, before spreading rapidly to the rest of the country. Finance for property development was traditionally found by a *"tour de table"* or syndicate of private investors, each providing a proportion of the finance needed as a short term measure, in return for a share in the trading profits of the completed building. The eventual occupiers would purchase the premises on the basis of occupation on its completion. For many French companies, this tradition died hard and indeed has by no means disappeared completely. Speculative property development has become, however, the more usual pattern for office building in France since the early 1970s, following the example set by British property interests.

Leasing agreements in France are somewhat different than those found elsewhere and their form does have an influence on the development process. A *'crédit bail'* contract was introduced in 1967 for commercial property. It involved a fifteen year lease during which the occupier pays an annuity representing capital repayments and interest. After this period, the occupier may acceed to the freehold for a peppercorn rent. Essentially this is a form of 'hire-purchase', but is usually financed by lending institutions (SICOMI's) which were set up specifically for this purpose in 1967. Although some office development is developed on this pattern, far more frequently the development companies use the *'bail commercial'* pattern under which nine year leases are granted with rent reviews after three and six years. This type of lease originated in 1953 as the statutory commercial lease. Rent is indexed either annually or every third year, linked to one of the official indices of construction costs - usually that calculated by INSEE. At the end of nine years, the rent reverts, either upwards or downwards, to the prevailing market rental. Of particular note is the fact that tenants may opt out of the lease after either the third or the sixth year - a particularly tempting possibility if indexation of rent increases by construction costs have run ahead of market rental values, since cheaper accommodation may be obtained elsewhere in this market. Speculative office developers, therefore, had to learn to work within the constraints of this system of leasehold and could not invest on the basis of expected high returns from periodic rent reviews imposed by the landlord on the tenant. This protection of the commercial tenant was reinforced by the fact that the tenant generally has the automatic right of renewal of the lease after nine years. The balance between landlord and tenant is therefore, rather more even than in Britain during its office 'booms' when property companies could dictate terms of tenancy.

A further restraint on the market stems from the taxation systems as applied to dealings in property. During the first five years' life of any building, its sale would attract TVA (value added tax) at the rate of 17.6% but such tax would generally be recoverable against future costs of operation. After five years, however, a 'registration' tax or duty is levied on the transaction. This is generally at a rate of 17.1%, to include other minor taxes, though it should be noted that there are slight regional variations in this level of duty. Considering the brisk trading in office buildings that takes place elsewhere, it can be seen that this level of tax is a strong brake on the market in established property and may make France an unattractive market for some property investors seeking to acquire established property rather than new developments.

Other fiscal measures have a bearing on the office development market and the response of developers. Two specific measures have had their effect, one directly as a restraint on development and the other as an indirect encouragement to commercial property development. The restraining measure was passed in 1978 and is known as the *Loi Galley*. Besides incorporating a land development tax, this legislation gave public authorities the right to the acquisition of properties more than ten years old at the agreed market price in order to permit the assembly of sites for public facilities. More directly constraining, however, were the limits on building densities on the redevelopment of a site. Plot ratios in urban areas throughout France were limited by the *Loi Galley* to 1.0, whilst a slightly higher density of 1.5 was permitted for Paris. Development of higher density attracted a betterment levy. The second, more recent measure is the *Loi Quillot*, passed by the socialist government in 1982 and giving considerable security of tenure to residential tenants. Its effect on the commercial sector has been to deflect domestic investment finance away from its traditional area of residential property development to other areas. Although it is too early to quantify its impact, it seems likely that the French institutions will begin to compete more vigorously in the industrial and office sectors of the property market. In itself and combined with the more general decentralisation policies adopted by the Mitterand government, it is fair to speculate that this may lead to the development of a more sophisticated office market in provincial France, the lack of which has been a notable feature of the French office development process.

LACK OF A PROVINCIAL OFFICE MARKET

The historical dominance of Paris over all sectors of French commerce, as well as art and culture, has led to a considerable imbalance in terms of office location. Some 50% of the office space in France is within the Paris region, compared to the 37% of the total office space in England alone which was situated in Greater London in 1982. The headquarters of French companies are usually to be found in Paris and this is particularly noticeable in the case of the financial commercial sectors - 96% of French banks have their headquarters offices in Paris and over 40% of those employed in banking, insurance and business services in France are employed in the Ile-de-France region. This concentration has led to the growth of the tertiary employment sector in the capital, with approximately two thirds of employees now employed in the service sector generally. In analysing this concentration of office activities in the capital, Tuppen notes that 31% of French office workers work in Paris, compared to only 3% in Lyons, the largest provincial city. (Tuppen, 1983). This excessive concentration has led to a remarkably poorly developed provincial office market, certainly when compared either to the UK or to the rather different case of the Federal Republic of Germany, which has a number of distinct office markets, including Dusseldorf, Frankfurt and Munich. Whilst companies employing more than ten persons occupy 27 million square metres of offices in the Paris region, the twelve largest cities in France house a total of only 10 million square metres of offices of companies of this size. A comparison of a similar nature shows that of large office premises (i.e. over 300 square metres and therefore employing 15-20 persons), the GLC had in 1982 only 20.4 million square metres, compared to a total for England of 55.3 million square metres. Since this total excludes Wales, Scotland and Northern Ireland, it would seem clear that the provincial office market is very much stronger in the UK than it is in France. Table 6.1 summarises the general pattern of office employment in provincial cities in the mid-1970s, demonstrating the fundamental imbalance described above.

The weak provincial office market stems in part from the strong tradition of owner-occupation considered earlier. Inroads into this were first made in Paris in the 1960s, but even by the mid-1970s, major provincial centres had changed little in this respect. For instance, Strasbourg may have been expected to be a likely recipient of new offices built speculatively to feed off its growing European status

Table 6.1: Major Centres of Office Employment in France

Agglomeration	Number of office employees 1975	Proportion of total office workers in France (%)
Paris	1,538,660	30.9
Lyon	150,820	3.0
Marseille	119,420	2.4
Lille	104,300	2.1
Bordeaux	76,405	1.5
Toulouse	64,310	1.3
Nantes	55,230	1.1
Strasbourg	46,880	0.9
Rouen	46,090	0.9
Grenoble	44,720	0.9
Nice	44,135	0.9

Source: INSEE, *Recensement* (1975). (from Tuppen 1983).

particularly after the enlargement of the EEC in 1972. A review of the office market there in 1974 concluded that

> "there are definite signs that no letting market exists in Strasbourg - one of the only modern office buildings in the city stood empty for several years without being let, and the company which is dealing with the Place des Halles scheme (a large-scale redevelopment of mixed areas, but including 28,750 square metres of offices) is now its only tenant occupying one floor. British investors will need a great deal more evidence that there is a move away from the concept of owner-occupation before investing in the city."
> (Estates Gazette, 14 September 1974, Vol.231, p.1231).

By 1980, more speculative office development had been created - 100,500 square metres in all, but this was still a small amount by the standards of most established office markets.

Whilst it is true that other cities, especially Lille, Lyons and Marseilles, have demonstrated some potential for profitable speculative office development, inroads into the predominant position of Paris have been negligible. Central Government attempted in 1965 to redress the provincial-Paris imbalance by the designation of the *métropoles d'équilibre*, intended to decentralise much of the growth of Paris, especially in the tertiary sectors into a series of major provincial growth poles. (See House, 1978, or Tuppen (1982) for a more detailed appraisal of this policy).

Such a policy may have been expected to stimulate office jobs and therefore an active office market in the provincial centres concerned. It was set against a background of 25% of the new jobs in France being created in the Paris region between 1962 and 1969, with the consequent tertiary growth in the capital. From 1967, DATAR (*Délégation a l'Aménagement du Territoire*) - the French government regional planning agency - instituted investment grants for firms either establishing or transferring administrative, research or development activities to fifteen named cities (including the twelve *métropoles d'équilibre*). In 1969, new office development of over 1000 square metres required sanction of an inter-ministerial committee, whilst two years earlier, in 1967, occupiers proposing to move into office units in excess of 3000 square metres were required to obtain permission from the decentralisation committee (*Comité de Décentralisation*). The structural constraints on the office development process in Paris were therefore instituted early and certainly prior to the boom in speculative office development instigated by British interests in the years around 1970.

The effect of such controls on the national office market was minimal. This was particularly unfortunate since some of them, such as the occupation permit, were a sensible attempt at determining which offices had a real case for a Paris location, as opposed to those drawn there by tradition more than by need. Between 1968 and 1971, however, only 28 service concerns attempted to avail themselves of the tertiary investment grants (Cameron, undated), whilst the demand for offices in Paris was being acceded to at an increasing rate. In 1967, 325,000 square metres of offices were given approval in the Paris region, but by 1971, this had risen to 1.435 million, most of which was speculative development.

In summary, therefore, France offers little in the nature of a provincial office market and notwithstanding government efforts, few provincial cities have attracted large numbers of offices. Instead, the French office market is still to all intents and purposes, the Paris office market. It remains to be seen, however, whether decentralisation measures adopted by the present socialist government in France have a marked impression on this pattern of office investment in the country as a whole.

REGIONAL PLANNING AND OFFICE DEVELOPMENT IN THE ILE-DE-FRANCE

If a national redistribution of the office development has foundered on the rocks of French pre-occupation with Paris, there has been a degree of success in directing the office market within the capital itself. Indeed, as will become evident, rigorous controls have not only restricted office development in areas which had seen too much activity, but by the 1980s were succeeding in channeling investment in offices to very specific locations within the region. At a regional scale, therefore, the French government has shown that office development can be planned on rational principles and that the office development industry need not shape the pattern of offices in creating and reinforcing the location of offices and thus determining its own market.

The regional plan for Paris has been reviewed extensively elsewhere (see for instance, Moseley, (1980) and Burtenshaw et al (1981)) and all its details need not concern us in this discussion. On the other hand, its underlying principles and provisions are important to our understanding of the constraints on the development process. The original plan *(Schéma Directeur d'Aménagement et d'Urbanisme de la Région Parisienne)* published in 1965 included the building of eight new towns and the designation of nine suburban growth poles *(poles réstructurateurs)*. Much of the anticipated growth of the region was to be directed into these new centres in order to reduce the growing imbalance between place of work and place of residence in the region and to endow the hitherto impoverished suburbs of Paris with new centres to act as nodes for services and employment. By 1969 the number of new towns had been reduced to five - Cergy-Pontoise, St. Quentin-en-Yvelines, Evry, Melun-Senart and Marne-la-Vallée. Their target size had been reduced from an anticipated range of 500,000 - 1 million population in 1965 to 200-300,000 by 1976, in line with the downturn in the demographic growth forecast for the region as a whole. The suburban growth poles vary greatly in character from established suburban centres such as St. Denis and Versailles to rather extended *pôles* such as Le Bourget-Roissy and Orly-Rungis (Moseley 1980). The early growth of the suburban growth poles was very varied, reflecting the varied background of the centres themselves. Table 6.2 shows their demographic growth and the growth in service employment between 1968 and 1975 as they became established.

Certainly their early growth in service employment was encouraging and was in line with the general desire to provide employment in closer proximity to place of residence. At the same time, however, it will be evident that few of them had grown to be office centres as such although those associated with the international airports of Orly and Charles de Gaulle (Roissy) could be claimed to have had some success. On the other hand Créteil and to a greater extent La Défense had attracted offices in some quantity. La Défense is however, a special case since it was conceived in 1955 specifically as an office centre and as will become apparent below, its role in the Paris office market has been pivotal and crucial to the overall success of planning measures. The suburban growth poles had, therefore, some effect on the pattern of service employment, including that in the office sector. Only La Défense, however, rivalled Paris itself as an area for large scale office

Table 6.2: Population and Service Employment Growth in Paris Suburban Growth Poles 1968-75

	Population 1975	% growth 1968-75	Service Employment 1975	% growth 1968-75
Orly-Rungis	166,500	+3.8	63,600	+79.7
Vélizy	22,600	+45.8	7,200	+105.7
Versailles	118,900	+13.2	40,100	+25.3
La Défense	185,000	-0.7	76,900	+89.4
Saint-Denis	91,100	-7.9	25,600	+3.6
Bobigny	43,100	+9.4	12,400	+93.8
Le Bourget-Roissy	226,800	+17.0	5,200	+97.7
Rosny	35,800	+16.6	6,900	+102.9
Créteil	59,000	+19.9	20,200	+169
	953,000	+8.4	304,900	+69.5

Source: Moseley (1980)

development demanded by major domestic and international companies.

Whilst the suburban growth poles have been planned to be a location for service employment, including offices, and have met with a degree of success in this aim, it is the new towns which have been more specifically favoured by a variety of regional planning and control initiatives.

Both the granting of *agréments* and the *redevance* or development tax have been structured to favour the new towns. Since 1974, of the ceiling of 900,000 square metres per annum for entire Paris region, of which 20% was to be located in the five new towns. A similar advantage has been built into the *redevance* system since offices in new towns attract a rate below that for other sectors in the region. Initial development in the new towns suggested some success in these policies, with 760,000 square metres built in the new towns in the period 1971-1976, although this combined total for all the new towns was below the 830,000 square metres constructed at La Defense in the same period (Moseley 1980).

The situation by 1981 however, had improved markedly, as a detailed appraisal of the office market indicates. In that year, the new towns attracted 17.5% of the regional total of offices and received 30% of the *agréments* for the region, amounting to permissions for 193,000 square metres (IAURIF 1982). On the other hand, the distribution of the new office space between the new towns was very uneven, reflecting the more general east-west imbalance in office activity which has been a characteristic of the Paris region. In addition, not all the *agréments* resulted in offices actually constructed (see Chapter 9).

The final element in the infrastructure of planning and fiscal restraints on the office development came in 1977 with the *Schéma Directeur d'Aménagement et d'Urbanisme de la Ville de Paris* (see APUR 1980, for a detailed presentation of this plan). This plan set out policies for the central *Département de Paris*, or Paris *intra-muros* and considered the main problems to be faced by the city. Its aims included the limitation of employment growth, the combatting of an increasing structural imbalance within the city and the prevention of the geographical separation of the various functions in the city (Bateman and Burtenshaw 1983). Whilst its policy for office development was not one of complete restraint it was in general very restrictive. Faced with a loss of manufacturing jobs of 12,000 per year between 1968 and 1975 and a growth in service employment of 19% - 500,000 jobs - in the same period, it was not surprising that there was no strong desire to allow unchecked office growth. On the other hand, it was recognised that Paris was increasing in importance as a major financial centre, with a 29% increase in employment in this sector between 1968 and 1975. Thus in the established financial and commercial districts of central Paris, modernisation of offices and their replacement are permitted, but no increase in employment densities is allowed. Elsewhere, and particularly in the east of the city, specific encouragement had been given to office development since the 1960s, but such policies were aimed at redressing the east-west imbalance of the city and thus favoured areas such as the Gare de Lyon-Bercy quarter in the east of the city. The 1977 Plan, whilst not abandoning such policies, which had also been supported by lower rates of *redevance* since 1971, re-emphasised the need to preserve the residential function of the city in the face of increasing dominance of commercial land use. In keeping with this policy, permission is unlikely to be given for office development in cases where it would involve the displacement of residents from the city.

The framework for considerable direction of the market thus was complete, but as discussed in Chapter Four, this is no guarantee that control would necessarily be successful. In Paris however, there was a strong popular feeling that high-rise office development should no longer be permitted in the city centre, especially after the 56-storeyed Tour Maine-Montparnasse had disfigured its skyline. In a very general sense therefore, there was the political will to restrict offices in central Paris, although no-one wanted to see it lose its international status. It is the latter sentiment which questions the political willingness to deconcentrate offices into the provincial cities of France, and it is not yet possible to ascertain the extent to which the post-1981 government has demonstrated its political commitment in that direction. The office development industry had to work within the constraints of the planning system as well as challenge much of the established practices of development. In Brussels, the office development industry won at least the early rounds in the process of development in that city, but in Paris, office developers were allowed no such period of free rein and were soon kept in very close check.

THE RESPONSE OF THE DEVELOPERS

The move by British property companies into France in the 1970s was certainly important, but was less spectacular than that into Belgium. Nonetheless, some early pioneers such as the Heron Corporation were to have a strong impact and on the property sector of Paris in particular. That company was involved in the development of the Perisud scheme at the strategic junction of the Boulevard Périphérique and the A6 *(Autoroute du Sud)*, involving 23,224 square metres (250,000 square feet) of offices jointly financed with a second company, Ziron European Properties BV. It also acquired a majority interest in the vacated *Figaro* building and adjoining property at *Rond Point des Champs-Elysées*, probably one of the peak land values in Paris and symbolising the confidence of British property interests at that time. This building was eventually renovated as a prestige office and retail development (3,100 square metres of retailing and 9,500 square metres of offices) in a joint venture involving Heron, UAP (a French insurance company) and Keyser Ullman, merchant bankers, and completed in 1980. (Estates Gazette, 27 September 1980). Until 1974, there was a steady stream of British investment involving both prime sites and some somewhat speculative secondary sites. Amongst the former were developments let to French financial institutions, such as the 1,350 square metres of offices at 51, Ave. Kleber, let to the Banque Nationale de Paris by the E. Alec Colman Group and Sun Alliance and London Insurance Group's development of 5, Rue de la Bourse to produce 1,100 square metres of office space in the heart of the financial quarter and close to the Bourse and let to the Banque de d'Union Parisienne (Estates Gazette, 20 July 1974).

More peripheral sites were also developed or acquired during this period of establishment by British interests. For instance, a consortium headed by London and Overseas Property Investment Co. acquired part of Rondpoint 93 from a French bank, Banque Paribas, a 26,000 square metre development in inner suburban Montreuil. The development was apparently well placed with integrated bus and *métro* facilities, together with retailing and extensive car-parking. The finance of £15 million was raised by a 40% contribution from a consortium of British banks and finance companies, with the remainder coming from French sources. (Estates Gazette, 24 November 1973). This investment, however, illustrated many of the difficulties of the Paris market and especially the bifurcated nature of the market between east and west Paris. Montreuil in the east was not a traditional location for office activities and the British investors sold the development to a consortium of French banks, including Banque Paribas, at a loss after four years of failure to find a tenant. The continuing low level of demand for offices in eastern inner Paris meant that the building continued to lie vacant until 1983, ten years after its completion (Estates Gazette, 18 March 1983).

Investments were made in offices elsewhere in the Paris region during this period, including some in the new towns, such as the Commercial Union Properties financed development of 46,450 square metres (500,000 square feet) of offices in Evry. Campus sites were also developed, although not on a large scale, one good example being provided by the English Property Corporation's development in 1974 of La Boursidière, a 55,740 square metre complex of landscaped offices, strategically placed near Orly Airport to the south of Paris on the new A86 outer ring road of the city. (Estates Gazette, 23 November 1973). The tentative nature of this type of development was emphasised by the fact that in early 1976, only 35% of the development was let, although that figure had risen to 75% by the end of that year.

Involvement in the provincial market was very limited in the first half of the 1970s. The general lack of a strong market discussed above, was a major factor in this lack of activity. On the other hand, growing restrictions in the Paris region were persuading some investors to look elsewhere and by 1974, there were speculative developments in Lyons, Marseilles and Lille. Lille had twenty-one British-owned schemes, whilst in the centre of Lyons, the congestion of this traditional centre had led to the development of a major complex involving retailing, office and other central area functions. This development, le Part-Dieu, was part financed by British investors and included 325,150 square metres (3.5 million square feet) of office space (see Tuppen (1983) pp.294-296, for a detailed description and appraisal of this project).

Some office investment activity was in refurbished office schemes. This in itself indicates the degree to which planning strategies had limited development opportunities, since in a general sense, new purpose-built development is a far more attractive investment proposition, involving somewhat simpler procedures for a developer than the refurbishing of an existing building. An annual quota of office development of only 40,000 square metres in Paris itself (900,000 square metres in the region) however, was a strong incentive to look at alternative forms of office investment. Thus there was a steady though never spectacular activity in the refurbishment sector.

The early 1970s saw British property investors moving into Paris and investing widely in all types of offices, but the collapse of the British market in 1973-74 led to a degree of retrenchment just as it had from Brussels. It is interesting to note, however, that by 1976, rent levels were being maintained in Paris, against the trends seen elsewhere, buoyed up by the restriction on supply. This was balanced somewhat, however, by the very high stock of offices which were vacant in the Paris region, amounting to 536,000 square metres in central Paris, 1.17 million square metres in the inner suburbs and 300,000 square metres in the outer suburbs. (Financial Times, 20 October 1976). This period was therefore characterised by three particular trends. Firstly the sale of property by British investors in an attempt to balance their assets, secondly a growing interest by French institutions, who traditionally had invested mainly in the residential sector, leaving the prime sites to British, then Dutch and German funds, and thirdly some collapses of British companies. The first two trends were brought together by the sale of Chesterfield-Ronson's Opéra-St. Anne building, with 6,712 square metres of offices let to the Bank of Tokyo and the Banque Européenne de Tokyo as their French headquarters. It was sold for £11 million to VAP-Vie, a major French insurance group. The very publicised collapse of a British company indicated the frailty of the Paris market outside the centre and western suburbs since it involved a 40,000 square metre development on the Boulevard Périphérique at Porte la Villette, in the unfashionable north-eastern sector of the city. Amalgamated Investment and Property Co., the British developers involved were made bankrupt when their loan of £4 million, guaranteed by a UK bank, was called in by a consortium headed by Crédit Lyonnais (Estates Gazette, 13 November 1976). It was later offered for auction by the company's liquidators with a reserve price of Fr100 million, remarkably low since banking interests, led by Crédit Lyonnais had themselves advanced Fr200 million on the building. (Estates Gazette, 28 May 1977). This forced sale of a very large building at a low price was a fair indication of the need to consider the geographical diversity of the office market in the Paris region.

The market in 1976 was, therefore very weak and it is worth pausing to comment on the development of La Défense, considered in detail later in this chapter, in the context of the cyclical nature of the office market. At this time, it shared the general problems of the Paris market. Although favoured both by planning policies and by its geographical location in the western inner suburban *département* of Hauts-de-Seine, it faltered in this period, a situation which continued until a revival in the late 1970s.

Whilst there was in general an oversupply of offices from 1976 onwards, there was growing concern at the shortage of large office units. Even though there had been a 1.2 million square metre supply of office space (Richard Ellis survey 1978) at the end of 1977, annual take-up was running at around 700,000 square metres. Coupled with the fact that controls ensured that there was relatively little space available in central Paris (around 60,000 square metres in total), this led to an increasing shortage of office space for large companies. In turn, this greatly assisted developers at La Défense and to a lesser extent elsewhere in the region, since they were well placed to satisfy this demand for large office suites. Restrictions were then beginning to have a tangible effect deflecting demand away from the centre. Even the *Tour Manhattan* at La Défense which had been vacant since its acquisition by the property interests of the Kuwaiti Government in 1974 had let 30,000 square metres, some 50% of its total by the beginning of 1978.

The pattern of rents achieved in Paris in the mid-1970s was a reflection of the geographical variation in the market both in terms of its supply and its prestige. In central Paris, rents at the more prestigious locations were as high as Fr1500-1600 per square metre, at La Défense Fr600 per square metre, whilst the Bobigny and Rosny areas in the eastern suburbs could command only Fr250 per square metre. This level was explained by an oversupply of offices in the east as well as the intense competition for the relatively small amount of offices marketed in the centre. Neither could it be automatically assumed that permission would be easily granted for occupation of an office in the centre of the city. An example of the problems involved is provided by the British Post Office Staff Superannuation Fund's Louvre Business Centre, in the centre of Paris, on the Rue de Rivoli. The fund had become involved in this large-scale refurbishment in 1972, funding a British developer, but acquiring total control in 1975. The French Finance Ministry requested permission to occupy 13,000 square metres of the building at an annual rent of Fr13 million and not surprisingly, permission was not immediately forthcoming. After a delay of twelve months, however, permission was granted although presumably not without considerable deliberation.

Development prospects in the provinces were still very limited and some cities, such as Lille were suffering from an oversupply of office space. The general problem was one of disequilibrium particularly since within a restricted provincial office market, one development of relatively modest scale could satisfy a demand for a relatively long period of time. Thus Lille, which attracted British investment in the early 1970s had by 1979, 19,000 square metres of unlet offices, having let only 8000 square metres in 1978 with a further 20,000 square metres scheduled for completion in 1979.

The Paris office market, on the other hand, was undoubtedly responding to the controls imposed on development in the centre. Jones Lang Wootton's International Property Review in 1980 pointed to the shortage of space in central Paris, although regional supply at one million square metres was still above regional annual demand which at that time was running at 500,000 square metres. The decentralisation policy was itself affecting the market, with two particularly large relocations -

those of St. Gobain - Pont à Mousson which vacated 12,000 square metres in 1979 and of Rhone - Poulenc which vacated 36,000 square metres in 1981. In themselves, these moves left space to be filled in central Paris, but were especially significant in that the moves were to La Défense. Since both companies were major industrial corporations (to be nationalised by the socialist government in 1982), their example was important and strengthened the role of La Défense. Rhone-Poulenc sold its five central Paris buildings in 1980 to banking interests, 50% of whom were private Kuwaiti investors. They then arranged a leaseback transaction to continue occupying the buildings until their move to La Défense in 1982, after which the properties were renovated prior to re-letting (Estates Gazette, 2 February 1980). At the time, this was claimed to be one of the largest property transactions in Europe and had additional significance because all of the property involved was prime property of which there had been a shortage in central Paris. A similar move was made in mid-1982 when IBM moved its European headquarters to La Défense, vacating 20,000 square metres of prime office space between the Rue du Fauborg St. Honoré and Place de La Concorde. This space was taken by the *Caisse Centrale de Co-operation Economique*, a government body under the aegis of the Ministry of Agriculture, dealing with economic development in the Third World. The annual rent of Fr25 million paid by the French government for these offices - *le Cité du Retiro* - and the apparent ability to obtain an occupation permit is adequate comment on both the continuing attraction of central Paris even for government departments, and the total acceptance of an office letting market by the French establishment.

The continuing short supply of office space in central Paris had an impact on rent levels in secondary, though still central, locations by 1980. For instance rents in an eighteen month period to mid-1980 had increased 50% in the secondary locations of Gare de Lyon, Montparnasse and Boulevard Haussmann, to reach peaks of Fr900 per square metre per annum, compared to Fr1000-1400 in prime locations. (Financial Times, 6 June 1980).

Continuing shortages of space, including by 1981 medium-sized units of 300-500 square metres, meant that La Défense was benefitting considerably. Following a period of relative inactivity after 1974 referred to above, development had begun afresh to complete a planned 600,000 square metres of new space for the period 1981-85. The situation in 1981 was that there was virtually no new development in central Paris.

A major study of the office market carried out in 1982 enables us to assess the overall impact of the restrictions on office growth in the region and arrive at some conclusions concerning the achievement of the aims of the *schéma directeur* for the Paris region. (IAURIF, 1982).

In global terms, the Paris region saw the letting of

Table 6.3: Distribution of Newly Tenanted Office Space 1976-81 (000's square metres)

	1976		1977		1978		1979		1980		1981	
	Total	% Region	Total	% Region	Total	% Region	Total	% Region	Total	% Region	Total	% Region
Paris	135	31.9	131	32.4	77	19.5	96	30.8	107	35.5	69	22.4
Hautes de Seine	164	38.8	158	39.1	169	42.8	107	34.3	75	24.9	111	36.3
(La Defense)	(42)	(9.9)	(66)	(16.3)	(60)	(15.2)	(84)	(26.9)	(31)	(10.3)	(48)	(15.7)
Seine Saint-Denis	37	8.8	27	6.7	39	9.9	40	12.8	43	14.3	23	7.4
Val de Marne	45	10.6	56	13.9	63	16.0	35	11.2	40	13.3	30	9.8
Petite Couronne*	246	58.2	241	59.7	271	68.8	182	58.3	158	52.5	164	53.5
Grande Couronne** (except New Towns)	22	5.2	5	1.2	11	2.8	4	1.3	21	7.0	20	6.6
New Towns	20	4.7	27	6.7	35	8.9	30	9.6	15	5.0	54	17.5
Region Total	423	100	404	100	394	100	312	100	301	100	307	100

* *Departements of Hauts de Seine, Seine-Sant-Denis, Val de Marne*
** *Departements of Essonne, Yvelines, Seine et Oise.*

Source: IAURF 1982

307,000 square metres of office space in 1981, of which 38,000 square metres was refurbished space. This compared with a level of 301,000 square metres in the preceding year.

Table 6.3 shows the pattern of distribution of the office space brought into use in period 1976-1981. A number of trends are evident in the pattern of office lettings. Firstly over time, the progressive reduction in used office space was a general reflection of the weak economic situation. Secondly, although the trend is not sustained throughout the period, the decreasing proportion of let offices in central Paris is clear. The importance of the '*petite couronne*' of inner departments i.e. Hauts de Seine, Val de Marne and Seine Saint-Denis is clear, although equally important is the discrepancy within this zone with Hauts de Seine, which includes La Défense having a consistently high proportion of the used office space. In four out of the six years surveyed, La Défense and the department in which it is located, Hauts de Seine contained more than a half of all the offices occupied in one year. The new towns, on the other hand, have had rather more mixed fortunes, although a particularly large share of the offices let in 1981 were in the new towns. Once again, however, there is a geographical disparity in the distribution of these offices since 80% of the 54,000 square metres let in 1981 were in just two of the five new towns - Cergy-Pontoise (23,000 square metres) and St. Quentin-en-Yvelines (22,000 square metres).

Table 6.4 shows the pattern of office stocks in the period 1976-1982. The rapid decline of the region's stock of offices from the oversupply of the mid-1970s is clearly shown. The office stocks of Paris and the Hauts de Seine have dropped more rapidly than those of less popular areas, whilst that of the new towns has in absolute terms varied little. On the other hand, an examination of the percentage distribution of the office stock is revealing since it shows that the share of the available offices located in Paris itself fell from 20% in 1976 to 9.7% in 1981 and similarly that of Hauts-de-Seine fell from 39.7% to 17.1%. The remaining two departments of the *petite couronne*, Val de Marne and Seine Saint-Denis, had 27.9% of available stock in 1976, rising to 49.3% in 1981. Finally the share available in the new towns rose from 6.9% to 16.8%. The strong policy against Paris had certainly had an effect, although a less dramatic inroad had been made into the position of Hauts de Seine, virtually an extension of western inner Paris, and including the fashionable and high rent area of Neuilly between Paris and La Défense. The new towns meanwhile were beginning to attract tenants, although their stocks of offices were relatively high. In 1982, the stock of offices in Paris was scarcely sufficient to meet demand, whilst the departments of Hauts-de-Seine and Val de Marne had approximately one year's supply at current tenancy rates. Seine Saint-Denis in contrast had the equivalent of

Table 6.4: Office Stocks available in Ile-de-France region 1976-1982 (square metres)

	1976	1977	1978	1979	1980	1981	1982
PARIS	205,000	145,000	83,000	85,000	62,000	45,000	34,000
Hauts-de-Seine Seine-Saint-Denis Val-de-Marne	403,000 190,000 93,000	321,000 249,000 128,000	218,000 251,000 111,000	89,000 220,000 89,000	73,000 185,000 88,000	66,000 155,000 60,000	60,000 138,000 35,000
Total Inner Departments	692,000	698,000	580,000	398,000	346,000	281,000	233,000
*GRANDE COURONNE (excluding new towns)	47,000	38,000	39,000	36,000	39,000	25,000	25,000
VILLES NOUVELLES	70,000	68,000	56,000	66,000	46,000	48,000	59,000
TOTAL ILE-DE-FRANCE	1,014,000	949,000	758,000	585,000	493,000	399,000	351,000

**Essonne, Yvelines, Seine-et-Oise* and *Seine et Marne*

Source: (IAURIF 1982)

four years' supply of offices.

An analysis of the *agréments* granted in 1981 suggests a success in terms of controlling and restricting office growth. Table 6.5 indicates the level of office construction permitted in Central Paris, La Défense and the new towns in that year. Only 5.5% of the total space was accounted for by permissions granted for construction in Paris. The limit of 40,000 square metres fixed in 1975 for Paris was not reached and neither had it been in 1980 (30,900 square metres). These figures stand in remarkable contrast to the 295,000 square metres, including 205,000 square metres of speculative office space, granted for Paris in 1971. The permissions for development at La Défense mark the near-completion of the project. Thus the role of La Défense in the future will inevitably decline, possibly in favour of the new towns, which have consistently in the 1980s gained permits for over one third of the speculative office development in the region. In broader terms, the new towns may thus be expected to play their planned role in the regional plan with a balanced employment structure. Certainly by 1981 they had emerged as credible locations for offices and this was increasingly being recognised by major companies. The move of Crédit Agricole, a major French bank, to new headquarters at St. Quentin-en-Yvelines in 1982 was witness to this.

Table 6.5: Office Space Permitted in the Paris Region 1981 (square metres)

	Speculative	%	Owner-Occupied	%	Total	%
Paris	13,200	3.9	22,100	7.3	35,300	5.5
La Défense	90,700	26.8	116,000	38.1	206,700	32.2
New Towns	120,000	35.5	73,200	24.1	193,200	30.1
Rest of Region	114,100	33.8	92,700	30.5	206,800	32.2
Total	338,000	100	304,000	100	642,000	100

(Source: IAURIF, 1982)

The situation after 1981 was initially one of some uncertainty in that a change in political direction in France caused a faltering of investment in property. Whilst the interruption of a cycle by political events was not as marked as that in Hong Kong a year later, the change in political direction of the country was sufficient to cause institutional withdrawal from the French market, reducing their property holdings by two thirds (Financial Times, 5 February 1982). The fears which gave rise to that distrust of the property market were gradually allayed resulting in a degree of cautious re-investment, but unfortunately other factors, partly occasioned by political change and partly by external factors continued to make the property market a difficult one. Although high inflation should favour property investment, France had one of the highest rates of inflation in the EEC and this produced an unfavourable climate for foreign investment. This, combined with high unemployment and an effective devaluation of the franc following the 1981 elections discouraged large scale institutional investment in office property, particularly from overseas. By June 1982, the cautious return of the then nationalised domestic banks to office property investment was evident, although overseas investors were still cautious. Government action had however, reduced the ability of the domestic institutions to invest in property since pressure was being put directly on their funds for other purposes. Besides exerting this indirect influence on the property sector, the government was also able to influence it in other ways. One was *via* the *Loi Quillot* discussed above, whilst a second was through the introduction and operation of a personal wealth tax. The Financial Times (5 February 1982) went as far as to claim that "President Mitterand's new wealth tax is stepping up the pace of the silent revolution now taking place within the French property investment market." The tax was levied on personal wealth in excess of three millions francs and business assets of more than two million. Although there is no precise indication of the amount of property which came onto the market as a result of the tax, there is little doubt that it caused a widespread transfer of property ownership from private individuals to the institutions, including some in prime locations in central Paris.

LA DEFENSE - A DETAILED APPRAISAL

La Défense has figured prominently in our analysis of the French office sector. This is hardly surprising considering the scale of the project and its importance to the regional plan for Paris. In its cycle of development, it has been a microcosm of the national office market. As such, a more detailed analysis of this crucial project is appropriate.

La Defense itself stands at a strategic position within the monumental framework of Paris, lying on the continuation of the westward axis running from the Palais du Louvre, through Place de la Concorde, Champs Elysées and Etoile (Arc de Triomphe), to the Avenue de la Grande Armée in Neuilly. As such it had been the subject of a number of architectural schemes by architects, including Le Corbusier. In the mid-1950s, however, when a private developer constructed the *Centre des Industries et des Techniques*, as an exhibition and trade fair centre, the local authorities concerned, Courbevoie, Nanterre and Puteaux, proposed a scheme involving large scale office development at its core, but including parklands, residential development, and recreational facilities as well as necessary transport and service infrastructure. National government was at that time concerned with the problem of providing new office space in Paris and confirmed the feasibility of the project. Thus a development authority - *Etablissement Public pour l'Aménagement de la Region de la Défense* or *EPAD* - was set up in 1958 with a wide representation of local, regional and governmental interests.

The functions of EPAD were wide-ranging and included the acquisition of existing properties and rehousing of the population displaced by the project, the clearance of the land and provision of the complex infrastructure for the project. In all, the project covered 760 ha., but the principal office quarter (Zone A), covers some 160 ha. It was then an extraordinarily ambitious project involving the provision of 1.5 million square metres of offices, as well as 7,585 dwellings in Zone A and 11,726 dwellings in the lower density Zone B in Nanterre further to the west. Provision of schools and sport facilities as well as the new premises for the University of Paris X at Nanterre was also part of the overall plan. EPAD itself had responsibility for all the infrastructure excluding major public services such as the express *métro (RER)*, and the SNCF railway facilities. The office quarter itself consists of a pedestrian deck above which rise offices and residential dwellings whilst below are the transport and servicing infrastructure. EPAD maintains ownership and control of the latter, and in a somewhat unusual manner, sells the space above the infrastructure to office occupiers, or the ground area required for deep foundations to office developers. Occasionally the arrangement has been for a long lease of the space involved, an example being provided by the very large regional shopping centre (105,000 square metres), known as *Les Quartre Temps* which has been developed on an eight-five year lease, after which title reverts to the community.

The viability of the project was heavily dependent on a good communication network, as well as more generally on the restrictions placed on office development in Paris itself. The communications provided are in fact extremely impressive, with Line A of the *RER* running through La Défense from St. Germain-en-Laye to the west. From La Défense, the next *RER* station, some five minutes' journey time away, is Charles de Gaulle-Etoile in the centre of the business quarter of Paris. Other public transport provision include a station on the SNCF suburban railway system and a bus station accommodating services on eighteen routes. Further improvement in public transport focality will be made with the planned direct link of the SNCF line to the west of Paris serving Poissy and the new town of Cergy-Pontoise, to the *RER* (they are of compatible guages), giving direct access to La Défense from these centres. With the exception of a circular boulevard, the roads serving Zone A are generally built below the office complex, and include the A13 autoroute running directly below the central axis of La Défense, and other local access roads. Car parking is catered for with the building of 18,000 car parking places. It can be seen, therefore that its communications are indeed excellent and that to all intents and purposes it shares the public transport accessibility of a central city site with enhanced road access.

Its history, however, shows a pattern of development which is somewhat cyclical, but which can be seen as the product of two separate, but interrelated sets of factors. On the one hand, the operation of the 'normal' economic cycle has meant that the development of La Défense has had peaks and troughs of activity. On the other hand, the needs of office occupiers, coupled with the financial needs of EPAD determined particular phases of growth and types of office development. As Figure 6.1 shows, the pattern of growth over time has been uneven. Once basic infrastructure was provided, development was quite rapid, with 199,000 square metres of space being available in eight office towers by 1971, and 829,000 square metres completed in 20 buildings by 1975. The completion of the last office towers of this phase was over-shadowed by the very high stock of offices available in the Paris region at that time - approximately one million square metres in late 1974, some 25% of which was at La Défense. Not surprisingly, there followed a comparative lull in the development process with only an additional 39,500 square metres being completed by 1980. Office occupancy, however, did increase somewhat from the 72% rate of 1975 to 94% in 1980, a figure which was rapidly improving as companies were being affected by the increasingly successful controls on central Paris. Once the development cycle received this stimulus of increasing demand, further activity took place with 1.13 million square metres completed by 1983. By that time, the planned total of 1.55 million square metres by the end of the decade looked a rather more realistic figure than it had in the mid-1970s.

The economic pressures combined with other factors to dictate the type and style of office development at La Défense. The first generation buildings (see Figure 6.1) were built in pairs, planned in this form by EPAD. Each office tower was 100 metres high, covered a ground floor area of approximately 1,000 square metres and provided 26-28,000 square metres of office floor space. Thus the early towers such as Europe and Aquitaine, EDF-GDF, were distinctive additions to the skyline of Paris and as such provoked some public disquiet. The second generation of towers however, planned after 1972, demonstrated the problems of EPAD in maintaining its own financial viability. It permitted larger scale development, resulting in some very tall towers offering large amounts of floor space, such as the GAN tower (70,500 square metres) and the Tour Fiat (102,500 square metres), which on its completion in 1974 was the tallest office structure in Europe. The completion of this generation

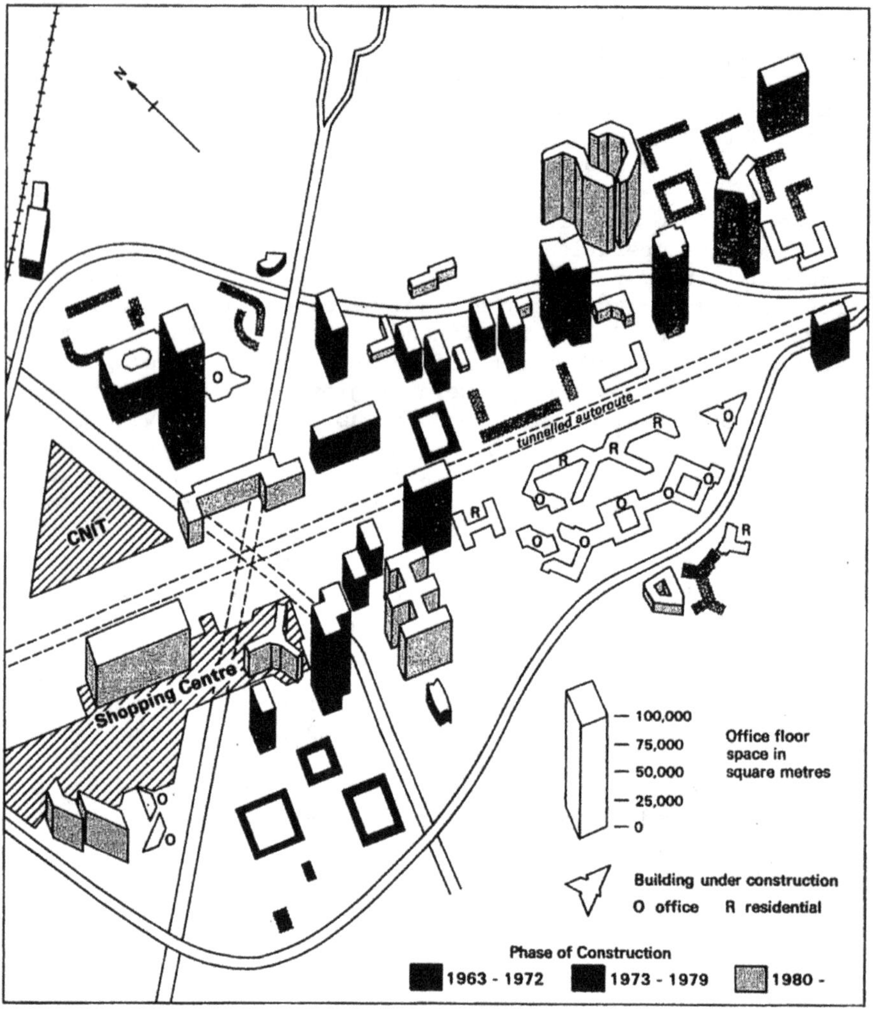

Fig. 6.1 Office Building Phases at La Défense

Table 6.6: Occupiers of Office Space at La Défense by Sectors of Activity in 1982

Activity	Company	Approx. occupied area	Total Space of sector (m^2)
Banking	Crédit Lyonnais	26,000	
	BFCE	1,000	
	Société Génerale	10,000	
	Citibank	12,000	
			49,000
Insurance	UAP	68,000	
	GAN	35,000	
	Winterthur	17,000	
	Préservatrice	25,000	
			145,000
Oil	Esso	30,000	
	Elf Aquitaine	126,000 (in 2 buildings)	
	Mobil	20,000	
	BP	10,000	
			186,000
Electronics	IBM	105,000 (in 5 buildings)	
	Rank Xerox	18,000	
			123,000
Automobile	Citroën	15,000	15,000
Chemicals	Saint Gobain	30,000	
	Dunlop	5,000	
	Pechiney	25,000	
	Hoescht Nobel	33,000	
	Ato-Chimie	9,500	
	Esso-Chimie	5,000	
	Rhone-Poulenc	27,500	
			135,000
Metallurgy	Usinor/Sacilor	35,000	35,000
Retailing	C & A	6,000	6,000
Energy	EDF	63,000	
	Framatome	50,000	
			113,000
Services	Transat	15,000	
	British Airways	5,000	
			20,000
Engineering	Technip	30,000	30,000

Source: Jones Lang Wootton.

of buildings did much to fuel the over-supply of offices at La Défense, which combined with other factors to produce a third generation of office buildings by the 1980s when development activity resumed. The new phase of buildings were energy-saving, depending more on natural rather than artificial lighting and were much lower-rise structures. The three phases of development and design combine to give La Défense a plot ratio of only 2.5, much lower than the immediate impression of clusters of office towers would suggest. However, the intervening circulation and landscaped areas mean that the density of building on the site as a whole is not especially high.

The success of La Défense was heavily dependent on two factors. One was the level of control exerted elsewhere in the region, and especially in Paris itself, and the second was the image of the project itself. It is true to say that its early image was not too attractive since major prestige occupiers were unlikely to forsake central Paris in large numbers for a vast building site. By the late 1970s, however, the project had reached sufficient maturity to attract such companies to La Défense. The list of occupiers, as Table 6.6 indicates, now includes some of the largest companies in France including the now nationalised Saint-Gobain and Rhone-Poulenc as well as important multi-nationals, such as IBM. As we have already seen, the movement of such companies to La Défense resulted in the only major office space to be marketed in central Paris in the last few years.

CONCLUSION

The recent history of office development stands as one of some considerable contrasts. La Défense, probably one of the most innovative office development projects in Europe, reflects the general fluctuations in the activity of the market, but also is a strong testimony to the success of strong regional control of the office market. French planners and government have been unwilling to accede to market pressures at the regional level. At a national level, however, the fact remains that despite an increase in office space in provincial centres, none rivals Paris as a location for major office activity and the provincial office market is undeveloped. It remains to be seen whether the decentralisation policies of the socialist government currently in power are sufficiently strong to promote an active office market.

In any event, a number of factors will need to change to persuade a return of foreign investment to the French office property market. Indeed recent evidence suggests a continuing disinvestment by foreign property interests (Bourdais, 1984). Before this trend can be reversed, the value of the French franc will have to recover somewhat from its low level of the last few years. This may make investment cheaper, but rent returns to foreign investors are weak compared to other investment opportunities such as the USA. The possibility of controls by a left-wing government which has already been very active in this field, is a further factor inhibiting investment. Certainly our major study area in the following chapter, the USA, offers marked contrasts with its recently strong currency and relative lack of controls on development.

Chapter Seven

NORTH AMERICAN OFFICE DEVELOPMENT

From a number of points of view, the office market in North America varies appreciably from that of Europe. Obviously it is divided between that of the USA and that of Canada, but there is also a greater geographical spread of office development activity than can be identified in European markets. In the USA, for instance, the traditional office market of the east coast, particularly that of New York, has been joined by cities further west and south such as the Texan cities riding initially on their oil economies and the Californian centres of San Fransisco and Los Angeles. More traditional centres of office development activity would include Chicago, with its often spectacular office developments, including five major office towers - Sears Tower, Standard Oil, John Hancock, One First National Plaza and Three First National Plaza - which between them offer nearly one million square metres of office space, around one ninth of the total downtown office market. (Financial Times, 9 March 1984). Each of these office markets has had its own pattern of activity, including periods of growth and recession which may not coincide in all cities. For instance, whilst the New York (Manhattan) market was in recession in the mid-1970s, causing a fall in general office development activity, that of Houston was booming, with the city's economy fired by its close association with and reliance on the oil industry. By the early 1980s, and the declining demand for oil, the Houston market was suffering from major problems of over-supply and development activity was being curtailed, whilst east coast markets such as New York and Boston were partaking in cautious recovery. In North America, therefore, there are considerable geographical differences in the development process.

On an intra-urban scale, there is also a highly differentiated market, to an extent unknown in Europe. Suburbanisation of office functions has given rise to separate office markets in many metropolitan centres. Major metropolitan centres such as New York, Chicago and Los Angeles have very significant suburban developments. Indeed in Los Angeles,

 DOI: 10.4324/9781003174622-7

only 38% of the Los Angeles-Orange County office stock is now situated in the downtown area of Los Angeles itself. Suburbanisation is not limited however, to major cities and any assessment of the office development process in North America must acknowledge this.

The process of office development in Canada has never been as dynamic as that of the USA, but major development has taken place in its metropolitan centres, such as Toronto and Vancouver. Calgary and Edmonton paralleled Houston in oil-related property development and have recently suffered many of the same problems of over-supply. The Canadian office market is worthy of close scrutiny, however, because it offers examples of the impact of government policies, both provincial and federal, on the development process. It will also become evident that Canadian finance has had an important part to play in promoting office development in US cities. Canadian development companies have been active in many cities, in some cases acquiring almost monopoly rights on future development potential.

This chapter will firstly review the differences in investment activity between North America and Europe before moving to a detailed examination of the recent trends in office development including a detailed appraisal of two US cities - Houston and New York. The Canadian pattern of office development is reviewed before concluding with a detailed analysis of the impact of government policies in Ottawa, its federal capital.

OVERSEAS INVESTMENT IN THE NORTH AMERICAN PROPERTY MARKET

The growth of office development in the USA and overseas participation in it during the 1970s and 1980s was very different from that of Europe, both from the point of view of financing and that of entrepreneurial activity. As indicated in Chapter Two, the emphasis of overseas property investment from Europe towards North America was marked in the late 1970s and early 1980s, as disillusionment spread concerning the profitability of the European market and particularly of its centre, Brussels. North America attracted investment not only from the UK and the rest of Europe, notably the Netherlands, but also from other countries such as Japan. Of course, one major difference between the office property market of Western Europe, as it existed in the early 1970s, and that of North America was the initial lack of a sophisticated market in the former, compared to its very obvious presence in the latter. Real estate and its investment in it, was by no means an an innovative concept in North America, which meant that overseas investors had to come to terms with a more competitive market. What is particularly important from a geographical point of view is that overseas interests were certainly successful so much so that they changed many American city centres irrevocably. These processes were, however, highly selective both geographically and temporally. Thus just as on the macro-scale, investment emphases switched from Europe to North America, so within the USA especially, certain urban centres were to enjoy a popularity for a period before going into a relative decline. Still others managed to maintain a continuous flow of investment in real estate, and that meant largely investment in offices, although we may note in passing that a great deal of investment capital has been channeled into North American retailing interests, particularly into regional shopping centres.

In the USA, in 1974, domestic institutional investors were for the first time beginning to show a real interest in property investment on the British pattern. US pension funds were beginning to be invested in property, although not on the scale of their counterparts in the UK.

The domestic investors in office property were in fact somewhat different from those in Europe. Broadly they were a) insurance companies, b) US Pension Funds, c) Real Estate Investment Trusts (REITs) d) property developers and e) syndicates of private individuals. Obviously insurance companies and pension funds operated much as they did in the UK, with the former requiring greater equity participation - as inflation in the 1970s began to erode returns from fixed interest loans. The REITs enjoyed special tax status being exempted from federal tax provided that 90% of income was distributed to shareholders. Private investors and commercial banks invested heavily in REITs in 1970-74, although the 1974 recession was to cause considerable difficulties for many of them (Estate Times Review 1979). The participation of private individuals in property investment is much more common in the USA than in Europe and frequently takes the form of a consortium drawn from the professions - lawyers, dentists, doctors etc. - who jointly own one office block. The proportion of total institutional investment in property, although accounting for a vast volume of investment funding has never been as high as that of their European counterparts. It is estimated that only 3% of institutional funds available for investment have found their way into property, owning in the process 4% of the commercial real estate market. These figures are a sharp contrast to the 25-30% of European institutional funds which are commonly invested in property.

There was an important difference between the USA and Canada in terms of overseas investment in property. In the USA, a free market was in operation and in a general sense it would even now be difficult to see a US government restricting overseas investment because of possible reciprocal restrictions elsewhere. In Canada, on the other hand, the Federal government passed in 1974, its Foreign Investment Review Act, which imposed a degree of limitation on overseas companies' investment in Canada. Whilst the legislation was actually aimed at preventing the so-called colonisation of Canada's natural resources it had an impact on the property sector. Not surprisingly, some companies managed to fulfil the conditions of the act sufficiently to invest in Canadian real estate, and investment from overseas continues, but in a general sense, more investors preferred the unfettered market of the USA, which in any event offered higher returns.

RECENT TRENDS IN OFFICE DEVELOPMENT

The situation in 1974 in the USA was not on the face of it particularly optimistic for office property investors. The traditional prime market of New York City was very depressed with 2.8 million square metres (30.1 million square feet) of vacant offices on the market. The recession of the early 1970s had combined with an oversupply generated by activity in the 1960s, including some very large projects such as the World Trade Centre in Downtown Manhattan. It should be noted that the Manhattan market was, and still is, divided into two zones, Downtown and Midtown Manhattan. The market centre, as reflected by rent levels has seen a general movement towards Midtown during the 1970s. In the mid-West, Chicago had a stock of 4.49 million square metres but a vacancy rate of 13%, underlining the poor state of the office market. On the other hand, other centres were seeing more activity. San Fransisco had a small supply of office (150,000 square metres), whilst Houston was seeing considerable activity with domestic investment in offices being joined by overseas interest, both European and Japanese. The development of Houston's office sector is considered later in this chapter.

At this point, it is worth considering why certain centres became important as recipients of investment, often from considerable distances, whilst others remain relatively untouched. Is it because they have particular investment potential linked to local industrial potential? Are there other factors generating an urgent demand for offices? The mutually supporting decisions of the property market often mean that local conditions may be a trigger mechanism, but once the process of development is begun, it continues because the institutions back each other's decision in the complex web of property deals which take place. Only when the message begins to filter through that local conditions are not generating the continuing demand does the process begin to wind down and even then, there may be considerable delays in its impact. The North American market exemplifies these processes during the 1970s and 80s particularly well.

An Economist Intelligence Unit survey of the USA market in 1974 gave information to potential investors concerning the state of the property market and indicated those cities with potential for a good return on investment. The cities were selected on the basis of population and economic growth. The Estates Gazette in publicising such a report (Estates Gazette, Vol.235, pp.37-39, 1975), acted as part of the information generating processes which sets the investment machine in motion and ensures the momentum so essential to its success. Five cities were seen as having investment potential - Los Angeles, Chicago, Washington DC., Houston and Dallas. Population growth in Washington of 39.6% (1960-70) was cited as evidence of market potential, compared to the 7.7% growth of New York during the same period. Chicago's population growth was only 12.1%, but it deserved attention because of its continuing economic significance as the largest city in the mid-West. Los Angeles was described as the major market on the west coast, although its sprawl ensured that the office market was actually highly sub-divided. Dallas and Houston were selected because of their importance for the oil industry and in the case of Houston, the location of the National Aeronautical and Space Agency (NASA)'s headquarters. Whilst one would not wish to deny the facts as presented in terms of demographic growth, there is a feeling of a self-fulfilling prophecy in the whole process of investment.

The need for good local knowledge was stressed and it is interesting to note that this was increasingly being made available by UK estate agents who were setting up offices in the US at this time. For the institutional investor operating from the UK at this time, perception of the market was determined by the property press, estate agents with knowledge of the market and the action of others. In a process where security of investment is so important, it is not surprising that there emerged a geographical concentration of investment in the office market. The process was therefore underway and by 1976 it was evident that some institutions had already invested heavily in the US. Amongst British investors were the Post Office Pension Fund, MEPC and Slough Estates. The National Coal Board (NCB) Pension Fund acquired Eastern Realty Investment Corporation, initially concentrating on purchasing property in the North Eastern states.

There is evidence that the US pension funds were beginning to invest in real estate, although not on the scale of either the British or Dutch funds. Few had more than 10-15% of their portfolio of investments in property and as noted earlier, most had very much less whilst it was not uncommon for UK funds to have 25-30% of their funds in property of various kinds.

A review of the US property market in 1979 (Estates Gazette, Vol.249, 10 March 1979) indicates the way in which the market was moving. The New York market was much better than in the mid-1970s, with rents improving and vacancy rate declining. It was not however attracting speculative office development on a large scale, largely because rents at $1.50 per square metre, were only just reaching the break-even point required to generate new investment. A number of companies, however, had elected to build their own headquarters offices in Midtown Manhattan, amongst them IBM, American Telephone and Telegraph and Philip Morris. Similarly the Chicago office supply situation was much improved following a continual over-supply dating from 1974 when 1.1 million square metres (12 million square feet) were added in one year. Houston continued as a centre where offices were growing rapidly - 375,000 square metres (4 million square feet) in 1979 - for which there was still a continuing demand.

Suburban developments, eg along Route 128 from Boston outside New York in Connecticut and northern New Jersey, were in demand. In the case of Connecticut, there was a growing demand for companies which had previously moved from New York and were requiring new support space. As a consequence, rents in Stamford, Connecticut were approaching those of Midtown Manhattan. The entry of Denver to the acceptable office market was also evident in this period. In 1978 it had 195,000 square metres (2.1 million square feet) under construction, but this was to be added to a supply of prime office space of only 836,000 square metres (9 million square feet). Denver then joined the growing yet still fairly restricted list of cities considered in the property reports and therefore seen as likely repositories for UK institutional investments.

The US office market saw considerable upheaval in October 1979, as a result of the Federal Reserve Bank's action in raising the discount rate by 1% as a part of a more general package of fiscal measures. This provides an excellent example of government activity affecting the investment process and thus indirectly affecting the urban development process. The growing economic recession was already resulting in decreasing demand for office space, whilst building costs continued to rise. On the other hand, as the Estate Times Review of the market (Estate Times Review 1979) indicates, investment in the US was still an attractive proposition.

The weak US dollar certainly made investment by overseas institutions very attractive and it is worth noting that investment other than in commercial property was taking place, with German investment primarily in land and French investment in residential property, besides a more generally based interest from Far Eastern investors. Thus, despite the set-back of October 1979, the US market was still favoured because of its long term prospects, backed by an underlying economic strength. Certainly in the case of UK institutions, wage inflation in the 1970s had left large funds to be invested and the US offered certain fiscal advantages over other investment markets. Real estate income, for instance, was largely, in some cases totally, free of US income tax. Thus the US was acting as a "tax shelter" for British funds, which responded with a steady flow of transatlantic investment. A result of the growing investment interest in the USA was the establishment of funds set up specifically to channel pension fund finance into US property, on a shared risk basis. Certainly by 1979 some of the larger pension funds had invested directly in the USA on a considerable scale. Somewhat to the later embarrassment of the mineworkers' union who are joint trustees of the fund, the NCB pension fund continued its investment in US real estate with the purchase in 1979 for $144 million of Continental Illinois Properties (whose portfolio included an interest in the Watergate Complex in Washington DC). Other funds of British nationalised industries, such as Electricity Supplies Nominees, British Gas Funds and Airways Fund all made direct purchases of property in the USA. On the other hand, the funds which enabled not only risks to be shared, but for them to be diversified geographically, were also attractive to many of these and other pension funds.

By 1979 five such funds existed and a scrutiny of some of them indicates the extent of British involvement in the US office property market. One such joint venture was established by the British-based Grovesnor Estates. Called Westcoast Freeholds, four major British pension funds held the remaining equity. Thus the British Rail Pension Funds, Airways Pension Fund, Boots Pension Fund and the British Broadcasting Corporation Pension Fund had by 1979 acquired a total of 74,311 square metres (800,000 square feet) of office and other commercial space with a strong geographical emphasis on California and other western states. Amongst its properties is the Southcentre Corporate Square, a 7.69 ha. site with office spaced leased to Boeing Computer Services and the Boeing Engineering Group. In 1980 it invested $20 million a 50% share in a 35,483 square metre (382,000 square feet) property in San Fransisco (Financial Times, 6 June 1980).

The American Property Trust was set up in 1974 with eleven unitholders, amongst whom were the British Steel Pension Fund, South Yorkshire County Council Staff Superannuation Fund, Post Office Staff Superannuation Fund, South of Scotland Electricity Board Fund, British Rail Fund, and the Electricity Supply Nominees. Most major unitholders were represented in the management committee of the fund which was initially chaired by the pension investment manager of British Steel Corporation. It is now advised by Richard Ellis and managed by a partner of the same company (Financial Times, 30 March 1984). Between 1974 and 1979, the American Property Trust raised £11.5 million for direct property investment in the USA. By 1984, the trust had assets of $250 million and its participants had grown in number to forty-nine, including major concerns such as Imperial Chemical Industries, Rank Xerox and Reed International. Its geographical distribution of ownership had shifted from a concentration in the south-east of the USA to include properties in Atlanta, Dallas, Houston, Washington, Chicago, Denver, Los Angeles and Kansas City. In early 1984 it acquired a 46,537 square metres (501,000 square feet) office development - the Three Penn Centre Plaza -

in Philadelphia for a sum of almost $50 million. It was also reported to have sold, after just two years' ownership, 208 South La Salle Street in Chicago to realise a trading profit of $28 million on its purchase price of $47 million. (Financial Times, 30 March 1984). The activities of property trusts such as the American Property Trust are therefore large in scale and they are sufficiently powerful to compete directly with domestic US investors.

The early 1980s saw some difficulties for investors in US offices. Construction costs were increasingly fuelled by the continued high interest rates. The *laissez-faire* attitudes of the 1970s were being replaced by increasing controls, such as the growing requirement for developers seeking building permits to carry out full environmental impact surveys (Financial Times, 6 June 1980). The so-called 'sunbelt cities' of the south, including Houston and Dallas, however, did not impose controls, ensuring continued development activity. As will be seen, however, oil-based Houston was soon to face its own problems of over-supply. The bases of UK property interests in the USA reflected the westward trend with MEPC, for instance moving its US headquarters from Minneapolis to Dallas (Investors Chronicle, 3 April 1981) and the Capital and Counties Property Group forming CapCount America, a wholly owned subsidiary of the British parent company, locating its offices in Atlanta.

The US government reduced somewhat the fiscal advantages of property investment in 1981, when it introduced a capital gains tax on real estate investment transactions, although there were loopholes in the legislation which diminished its impact (Financial Times, 15 May 1981). The early 1980s were in any event a more difficult time for property investment in the US. Whilst activity did not cease, it became more limited. The construction of 1980-81 created a market surplus in 1982-83. The economic recession was, moreover, resulting in a curtailed tenant demand. (Financial Times, 22 October 1982). A Richard Ellis report in 1982 suggested that two markets existed in the US. One comprised locations such as the suburbs of Houston, Denver and Dallas, together with the CBD of Atlanta and parts of Washington DC. All had seen considerable growth, with little constraint in the 'boom' years, culminating in a weak market suffering from over-supply. On the other hand, the stronger planning measures imposed in midtown Manhattan, Boston and San Fransisco had limited land availability for office development with the resulting stronger market, with less space available (quoted in Financial Times, 22 October 1982). By 1984 with a degree of general economic revival, the office property market was still differentiated between those cities which had considerable surpluses of offices dating from periods of excess activity and those which had begun to experience a diminution in office development in time for there to be a

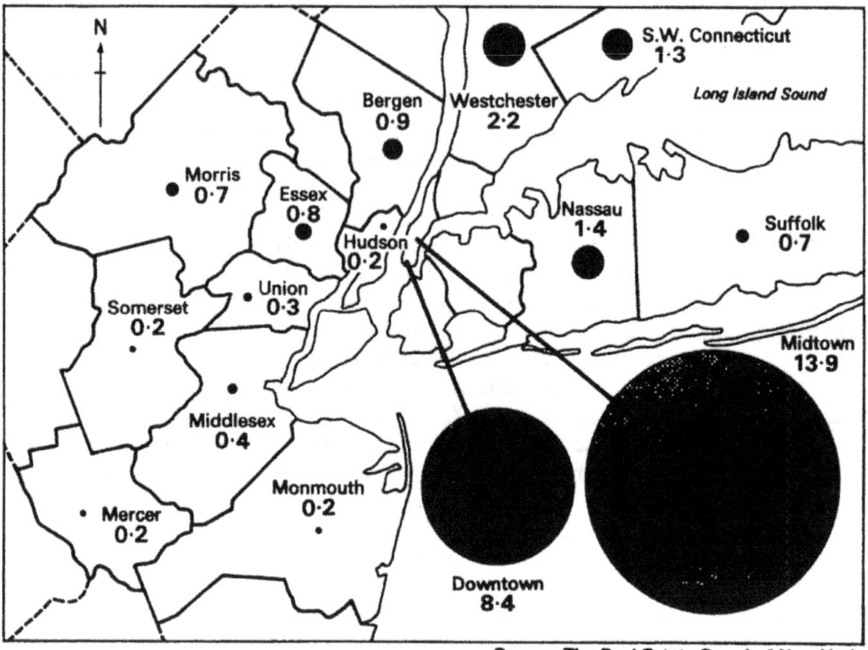

Source: The Real Estate Board of New York and Jones Lang Wootton 1981

Fig. 7.1 1981 Office Inventories of Counties within the New York Region in Millions of Square Metres

better balance between supply and demand. Two cities are considered here representing each of these two situations, with Houston providing an example of furious development activity resulting in oversupply and New York exemplifying a somewhat more cautious approach to office development, albeit on a larger scale.

NEW YORK - MANHATTAN AND SUBURBS

New York is in itself an internally differentiated market. It consists of the two office markets of Manhattan, Midtown and Downtown, complemented by a series of increasingly important suburban markets. Figure 7.1 shows the distribution of office space in 1981 in the New York region. There have been distinct differences in the patterns of development of each of these markets. The most important market is Midtown which has emerged since 1950 as the major centre for corporate headquarters, whilst the growing commercialisation of the south of Manhattan from the early twentieth century, created the Downtown market in the south of Manhattan Island.

The entire New York market has seen a considerable fluctuation in office development since the late 1960s. Between 1969 and 1975, some 4.6 million square metres (50 million square feet) of new office space was built in a surge of development activity which came to an abrupt end with the 1973-74 economic recession. From 1975 to 1979 only 8.5 million square feet of offices were added to the office stock. There is no doubt that the earlier period of activity produced an over-supply of office space in Manhattan, but this has gradually given way by the 1980s to a more balanced situation. A number of factors account for this change, besides the decrease in construction activity. Federal government support for the city's financial plight re-established domestic and international confidence in the city after its monetary crises of the mid-1970s. Recovery from the 1973 recession assisted demand for office space, sufficient for the renewed economic problems of the 1980s to have a diminished effect on demand. More specifically a measure was taken in 1978 to give New York a special status as an international insurance centre. The US government in creating a virtual insurance "free trade zone" in Manhattan was attempting to challenge the supremacy of the London insurance market and in so doing stimulated considerable activity, particularly in Downtown Manhattan. The growth of foreign banking, again encouraged by new central government regulations was a further stimulus to demand for office space with a rise in the number of foreign banks from 50 in 1972 to 250 by 1980. This increase in banking activity was especially concentrated in Midtown Manhattan, where Park Avenue has become the major axis of foreign banking.

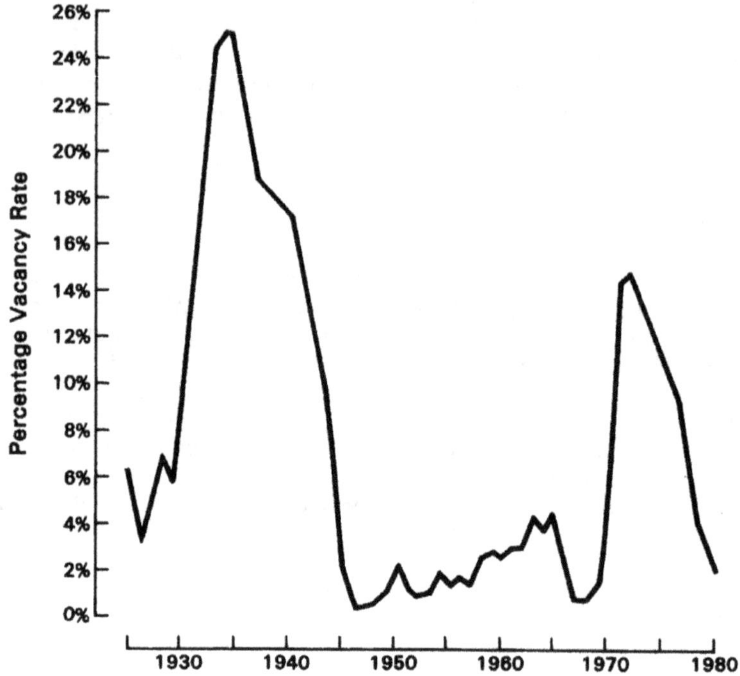

Source: The Real Estate Board of New York and Jones Lang Wootton 1981

Fig. 7.2 Office Vacancy Rates in Manhattan 1925 - 1980

By the late 1950s, Midtown Manhattan was the major office centre of New York. Large scale construction took place in the 1960s aided by revisions to the 1916 zoning laws, which permitted higher density development of offices provided that new developments incorporated ground level 'plazas'. (Jones Lang Wootton, 1981). Initial development concentrated on the Eastside of Midtown Manhattan. Whilst the whole of Midtown is bounded by 32nd and 60th streets on the south and north, and 3rd and 8th Avenues on the east and west, the half square mile core in the north east of that zone comprises Eastside. The activity of the 1960s culminated in the phenomenal growth of 1.77 million square metres (19 million square feet) in the period 1969-72. Expansion on that scale was particularly unfortunate in the light of economic recession and declining demand in the mid-1970s, although by the end of the decade recovery had taken place, with a vacancy rate of only 3.5% in Midtown Manhattan by 1979.

The original commercial core of the city was situated in the south of Manhattan, in the area now known as Downtown. Initially during the post-war period, increasing rent differentials between Downtown and Midtown reflected the growing popularity of the latter. The differential has been as high as $15 per square foot ($1.39 per square metre) in the 1970s, but rents now reach $45.50 per square foot ($4.18-$4.64 per square metre) in downtown, only slightly below the general level of $50 per square foot achieved in Midtown. The narrowing of the rent differentials underlines the revival of Downtown Manhattan as an office centre. Early century development of this sector produced major skyscrapers such as the Woolworth Building (1913) and the Equitable Building (1915). The pace of growth was such that control was felt to be necessary, resulting in the 1916 Zoning Law. The restriction on building densities produced the distinctive skyscraper office block with its stepped upper storeys and indeed construction of such buildings continued unabated until the economic depression of 1929. It was not until the 1960s that Downtown Manhattan was able to recover some of the development momentum which had shifted to Midtown. The building of the Chase Manhattan Bank's sixty-storey building and the World Trade Centre (approximately 1 million square metres) in the late 1960s assisted revival, only for the office market to collapse again in the mid-1970s. The overall vacancy rates for Manhattan are shown in Figure 7.2 and plainly demonstrate the vicissitudes of the market which affected the downtown.

The revival in the late 1970s was brought about by the increased importance of insurance and financial services. Downtown Manhattan already had four areas of office specialisation which could be distinguished - insurance, finance, government and shipping. The creation of an insurance "free trade zone" with tax advantages and a lifting of government limits on insurance risks carried, has stimulated office demand, and has included the building of the new New York Insurance Exchange. Further evidence of the continuing importance of downtown Manhattan is provided by the development of 37.23 ha (92 acres) of reclaimed land to the west of the World Trade Centre, including a 557,000 square metres (6 million square feet) World Financial Centre developed by the Canadian developers Olympia and York.

The Manhattan office market has therefore been extraordinarily resilient if one takes a long term view. Despite the growth in popularity of cities further west, attracting both corporate headquarters and investment finance, New York has reasserted its traditional attraction. At the same time, its suburban markets have shown very considerable development, originating in the 1950s and accelerating in importance during the next two decades particularly the 1970s.

The New York region has seen the establishment of four suburban office markets, which vary somewhat in their degree of concentration. They each have good transportation links to Manhattan, *via* both freeways and public transport systems. In addition, each is well connected to the regional and international airports of the region. The four suburban areas are Westchester, South-West Connecticut, Northern New Jersey and Long Island (see Figure 7.1). Together they have over approximately 10 million square metres of office floor space, a total exceeding that of downtown Manhattan's 8.36 million square metres (90 million square feet), and approaching midtown's 15.8 million square metres (150 million square feet). In aggregate therefore, the suburban market is extremely important and its status is such that although it grew in response to the need for expansion from Manhattan, it now has its own self-generating growth.

Westchester County, in New York State to the north-east of New York City has seen the development of 2.32 million square metres (25 million square feet) of office space. Its employment rose from 297,000 in 1960 to 369,000 by 1975. By 1981, it housed 500 office firms with 50 or more employees, including thirty-five corporate headquarters. Office development has been heavily concentrated on the network of freeways, especially the Cross Westchester Expressway, running across the southern part of the county. Some 80% of the office development lies within two miles of this expressway (Jones Lang Wootton, 1981). The centre of development is White Plains, where 250,000 are employed in offices with 668,000 square metres (7.2 million square feet) of office space, of which almost 300,000 square metres is in its central business area. Demand for office accommodation by both New York companies and others establishing in Westchester requires an annual development of approximately 100,000 square metres of new office floor space.

The south-west Connecticut market, like that of Westchester, is fairly compact in that although development has taken place in eight townships, most in the south west of

Fairfield Co., bordering Westchester Co., two townships, Stamford and Greenwich, account for most of the growth. In the 1970s, this amounted to 743,200 square metres (8 million square feet), with a particular concentration of major accountancy firms including the headquarters of Price Waterhouse in Stamford, the most important office centre. In 1981, there were 700,000 square metres of offices there, and 334,000 square metres in Greenwich, with a further 288,000 square metres elsewhere in Fairfield Co. Once again the freeway system has been responsible for orienting the development which in this case has followed the Interstate Highway 95, along the Connecticut coastline, the Merritt Parkway running parallel to it inland, or interconnecting highways.

In contrast to the relative concentrations of Westchester and S.W. Connecticut is the pattern of more dispersed development in Northern New Jersey. Here, the radiating highway corridors west of New York and Newark have proved attractive to office users and developers. In 1982, almost 4 million square metres of office space existed in the nine counties of northern New Jersey surrounding Newark. With an annual absorption rate of nearly 100,000 square metres (1 million square feet) in Bergen County, immediately across the Hudson River from Northern Manhattan, it is evident that this area has developed as an overspill centre for Manhattan "back office" operations such as computer and other technical support divisions of major companies. Elsewhere in New Jersey, growth has taken place based on good freeway access, particularly to New York itself.

The fourth suburban market in the New York region is Long Island, comprising Nassau and Suffolk Counties. The growth in the former was spectacular in the 1970s. In 1968 it had an office stock of only 315,859 square metres (3.4 million square feet), but by 1979 this had risen to almost one million square metres (10.5 million square feet). By 1981, there was an estimated office stock of 1.4 million square metres in Nassau Co. and a further 715,000 square metres in Suffolk Co. Whilst annual demand of 158,000 square metres (1.7 million square feet) in the early 1980s suggests a very bouyant market, it is also true that Long Island's communications with New York are less good than those of other suburban office centres. (Jones Lang Wootton, 1981).

It is plain that the world's largest office market of Manhattan with its 25 million square metres (250 million square feet) of floor space has now been complemented by very significant suburban growth centres which are sufficiently attractive in their own right to attract investment capital. As development opportunities in cities such as New York become more limited and increasingly expensive to acquire, suburban centres are viable alternatives in many north American cities, though not all cities have such a sophisticated suburban market as that of New York.

HOUSTON - OIL-INDUCED BOOM AND DECLINE

Suburban development has in proportional terms been of even greater importance to Houston, where the scale of office development in the 1970s was the largest of any North American city. Jones Lang Wootton's review of the Houston market in 1981 emphasised the continuing demand for office space, despite the unprecedented rates of office development in the 1970s running at the level of more than 400,000 square metres (4.5 million square feet) each year. By 1981, of its office stock of over 6.5 million square metres, one third dated from the period since 1975, most of it in suburban locations. A pause in developments in the downtown between 1975 and 1978 caused an increase in demand, resulting in rising rents before new activity began to fulfill the demand. By 1981, whilst the suburban office centres had vacancy rates of 6.2%, the down-town figure was only 2%. The suburban centres of note are the West Loop Galleria, with over 1.25 million square metres of office space and North Belt, in the vicinity of Houston Intercontinental Airport with nearly 0.5 million square metres.

The explanation for Houston's phenomenal growth lies in the importance of the oil industry to the city during the 1970s. Almost half of the employed population worked in the petro-leum refining or petrochemical industries, creating an economic stability unknown elsewhere. Rapid population growth - 40% between 1970 and 1979 to a total of 2.9 million - made it the fastest growing US city at that time. Other factors con-tributed to the remarkable office boom, not least the liberal local planning laws and minimal regulation of development in general. (Jones Lang Wootton, 1981).

The occupants of the new offices reflected the structure of the local economy with 40% of the space in 1981 being occupied by energy related companies, 21% by finance, insurance and real estate, 21% by miscellaneous service industries and 18% by other occupiers. Unfortunately its economic base which allowed it to weather the recession of the mid-1970s was vulnerable to the decline in demand for oil resulting from the recession of the early 1980s. Of the city's stock of 12.63 million square metres (136 million square feet) of offices, it was estimated in 1984 that 3.25 million-3.71 million square metres were vacant, producing a downtown vacancy rate of 13%, rising to 30% in some suburban areas. (Financial Times, 30 March 1984). Although the absorption rate was 750,000-850,000 square metres *per annum*, it was estimated that net take-up of space was only 275,000 square metres *per annum* because of the activity of office tenants consolidating their activities or up-grading their accommodation at a time of a weak market.

The boom period of the 1970s attracted foreign investment on a large scale. Amongst such investors was Cadillac Fair-view, a major Canadian developer. The activities of Canadian developers is further discussed below, but in the context of Houston, Cadillac Fairview's operation are of particular interest. The company became involved in the mid-1970s with Texas Eastern Corporation, an energy company based locally. Together they formed a joint venture company which owns a vast tract of land stretching from the downtown eastwards to Highway 59. At the centre of the city, it owns the Houston Centre, which stands on 17.5 ha. (43 acres) and consists of 283,000 square metres (3.05 million square feet) of office space in three separate blocks, together with the recent The Park in the Houston Centre, with a 27,869 square metres (300,000 square feet) shopping mall and an additional 63,822 square metres (687,000 square feet) of offices. Other features of the centre are a major hotel and the 1,300 space Houston Centre Garage (Estates Gazette, Vol.260, 14 November 1981). Spanning eastwards from this vast complex (worth $86.4 million at 1981 prices), the company acquired land on a considerable scale. It was reported as owning 33 blocks of streets running in a tract out to Highway 59, with 80% of the 300 ownerships originally involved being acquired in just ten days. This large scale land acquisition has given the Canadian developers control over a large area of land, for which major facilities such as a convention centre, as well as further commercial development were planned (Estates Gazette, Vol.260, 14 November 1981). This example of Canadian development initiative is merely the largest in a considerable collection of Canadian office developments, emphasising further the importance of international property investment. The following section illustrates this particular pattern of investment more closely.

CANADIAN INVESTMENT IN US OFFICE PROPERTY

Before examining the Canadian office market in detail, it is appropriate to consider the considerable influence which Canadian investors have had on the US property market. Already their influence on Houston and its development has been emphasised. Canadian influence has taken the form either of property purchase as a long term investment or the direct development of property as the prime motive. The mid-1970s saw the beginning of large scale Canadian activities. Foremost in the early dealings was the purchase in 1976 by Olympia and York of eight office blocks in Manhattan for $350 million. At that time, such a transaction was seen as unprecedented, particularly since the local New York market was languishing with over 2.75 million square metres (30 million square feet) of vacant office space. The value of the acquired property, however, increased fourfold by 1982 to be appraised at $1 billion. This particular property transaction was important since it emphasised the potential scale of Canadian operations in the USA. Certain parts of the USA proved to be particularly attractive to Canadian interests, notably Texas northwards to Denver Colorado and the west coast.

Denver particularly has been the target for investment with twenty four Canadian firms investing $1 billion there by 1982, culminating in Canadian ownership of more than 1214 ha. (3000 acres) of the city (Estates Gazette International Supplement, Toronto 1982). Cadillac Fairview, besides its Houston interests discussed above, has developed offices in Fort Worth and San Antonio as well as two major developments in Dallas and several west coast interests. On the west coast itself, Carma Developers have developed a 139,333 square metres (1.5 million square feet) project in Seattle, where Cadillac Fairview and CHG International have also jointly financed a 37,159 square metre (400,000 square feet) office development. Further examples of large-scale Canadian investment on the western seaboard are provided by very large scale developments in Los Angeles. Oxford Development have promoted a major office centre in the downtown area, comprising twin 52-storey towers, totally 120,769 square metres (2.6 million square feet). Still in progress is Cadillac Fairview's California Centre, a three tower project providing approximately 300,000 square metres (3.225 million square feet) of office space and scheduled for completion in 1991 (Estates Gazette 21 November 1981).

Not all Canadian investment has been entirely profitable. Indeed in the early 1980s Canadian property companies had to weather the reduction in demand as much as any other company. Some companies had to rationalise their activities considerably, a notable example being Daon which was forced in 1983 to dispose of two major San Fransisco building sites and withdraw its development headquarters to Vancouver from San Fransisco.

The Canadian influence on US cities is an excellent example of the power of mobile investment capital to shape and re-shape cities. The skylines of US cities both suburbs and downtown have been altered under the stimulus of a steady flow of foreign capital. Even in recession, the US has proved to be a magnet for foreign funds because of the lack of restrictions on either the inward flow of property-directed capital or too often, on the development process. An analysis of Canadian office development provides some marked contrasts.

OFFICE DEVELOPMENT IN CANADA

The Canadian office market provides fascinating examples of the varying impact of economic recession, and of both provincial government and federal government on the office development process. The cities which have seen major office activity are generally the provincial capitals, plus the federal capital

Ottawa. In Alberta, however, both Calgary and Edmonton have seen significant office growth, attributable to the general activity of the "oil corridor", which incorporates cities such as Houston, already discussed in this chapter. The graph of office development since 1974 (Figure 7.3A) shows clearly the pattern of office growth, including the overwhelming dominance of Toronto. The trends in growth of office rents shows a varied pattern (see Figure 7.3B). In most cities, the 1970s saw a steady increase in rental values, with a decline usually noticeable in the 1980s as the recession began to affect demand. Calgary, Montreal and to a lesser extent Ottawa, however, can all be seen to have had particularly marked declines in rental values in the early 1980s and the three cities will be discussed below to illustrate the effect on the office development process of the three factors of economic recession and, respectively, provincial and federal government policies.

Calgary - oil-induced growth and decline

The case of Calgary is interesting in that it shows the effect of economic recession on a city whose office stock rose very rapidly in the 1970s, from below one million square metres in 1975 to over three million square metres in 1984. Energy industry induced activity and confidence combined to create a property boom in the city shared only by Toronto and its growth was for different reasons. Vacancy rates have been rising rapidly as a declining demand for petroleum and petroleum products sapped economic confidence in the city. The result was a downtown vacancy rate of approximately 23% at the end of 1983, and an overall rate of 25%, or just in excess of 750,000 square metres of vacant space. Naturally, such a drop in demand for space had its effect on rent levels, with prime rates in 1984 for downtown offices at 50% of equivalent levels in 1981. (A. E. LePage Ltd., 1983). Positive developments for the return of office development confidence include the continuing construction of the city's rapid transit system and the subsequent stimulus of office park development in sectors such as the north-eastern suburbs of the city, which the system will reach in 1985. Nonetheless, elsewhere in the city, the general pattern has been for building projects to be held back and planned developments totalling over 1.5 million square metres scheduled for commencement in the period 1982-1986 have been postponed.

Montreal - shifting metropolitan balance

The effect of provincial government policies on the development process are shown by office development trends in Montreal and Toronto, shown in Figure 7.4. Yeates (1975) has noted the general movement of economic activity and growth towards the western end of Canada's metropolitan corridor stretching from Quebec City in the east to Sarnia in the west. Thus the impact of most recent provincial government policies has to be

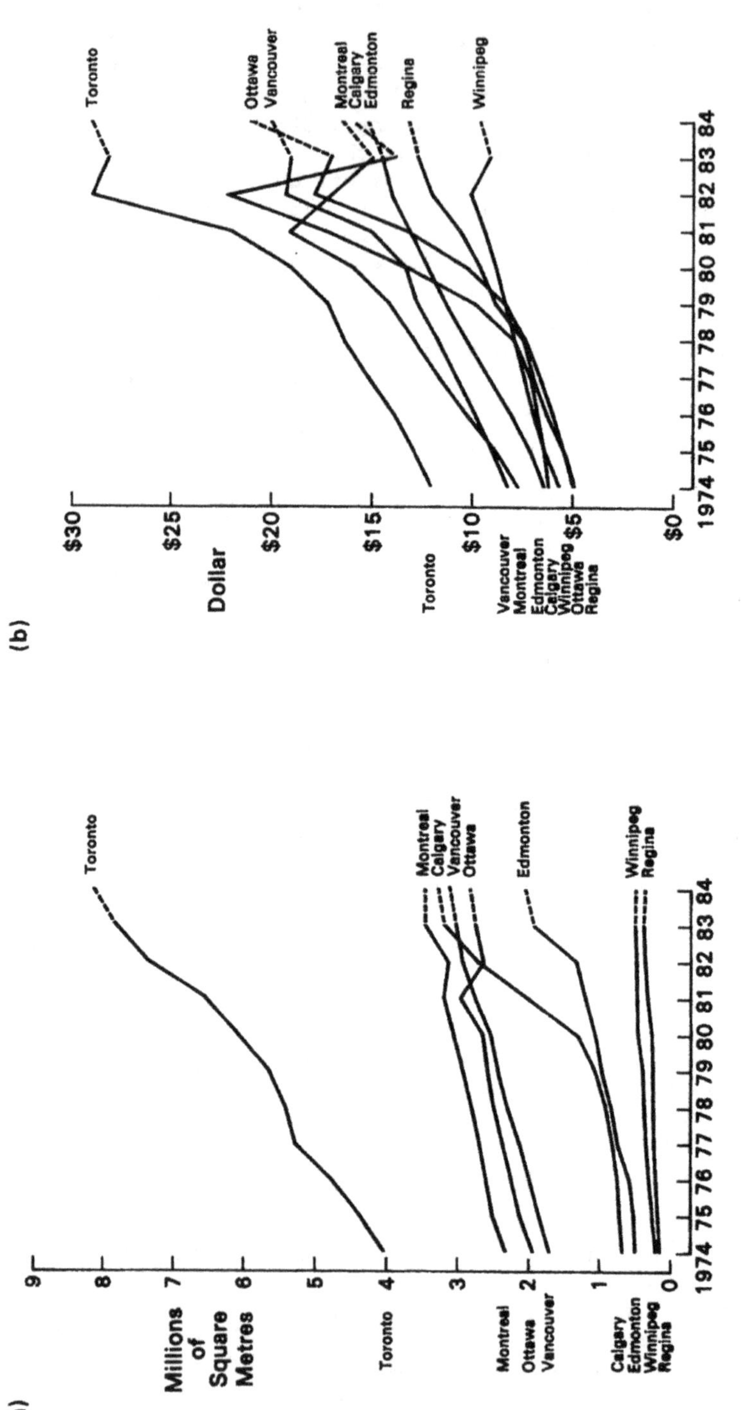

Source: Canadian Real Estate 1984 (A.E. LePage)

Fig. 7.3 Canada (a) National summary of office space in cities over 500,000 (b) Office rental rates – major markets

set against a general background of a shift of emphasis from east to west. The Quebec government, however, did little in the 1970s to instill investment confidence in the office sector of Montreal. In particular, the passing of Bill 101 in 1972, to encourage the use of French as the principal business language of the province, caused a physical movement of offices from Montreal. A notable example was the decision in 1978 by Sun Life of Canada to move its Corporate Headquarters to Toronto following the passing of the French language legislation.

Ottawa - Federal Government Policy Effects

Whilst provincial government legislation has had an indirect effect on office development in Montreal, a more direct effect of public policy has been felt in the federal capital, Ottawa, but in this case it is federal government policy which has occasioned major changes. A policy was adopted in 1969 to reorganise both the structure of government office holdings and their location within the Ottawa region. At that time, the Canadian government occupied 576,000 square metres (6.2 million square feet) of crown-owned (i.e. federal government) office space, but also occupied the not inconsiderable total of 411,500 square metres (4.43 million square feet) of rented space. The programme adopted by the Canadian government was to have a large impact on the latter since one of its basic aims was to reduce its dependence on the private sector and move towards providing a larger proportion of its own office requirements. A second and parallel aim was to give a reality to the Ottawa-Hull region as the National Capital Region. The key issue behind this aim was that the region straddled the Quebec-Ontario boundary of the Ottawa River. Most of the federal government offices were in Ottawa on the Ontario side of the river rather than in Hull, but the newly adopted Federal Office Relocation Program (FORP) foresaw a redressing of the imbalance. The policy adopted and pursued during the 1970s by the Canadian government under this program was for 75% of federal employees to be located on the Ontario side and 25% in Quebec by 1985. Other facets of the policy included the demolition of obsolete temporary buildings and a general consolidation of government departments where appropriate.

The potential impact of such a policy can be gauged by the fact that in 1969, 58% of the 613,000 square metres (6.6 million square feet) of offices in private ownership in the central area of Ottawa were leased to central government and 80% of all occupied office space was in federal government use. Although the plans for decentralisation of federal employees from the central area dated back to Greber's Plan for the National Capital in 1950, to include suburban campus developments (Nader 1976), the fact remained that the private office sector was still heavily dependent on the continued presence of federal government departments. The federal government was keenly aware of the likely effect of such a

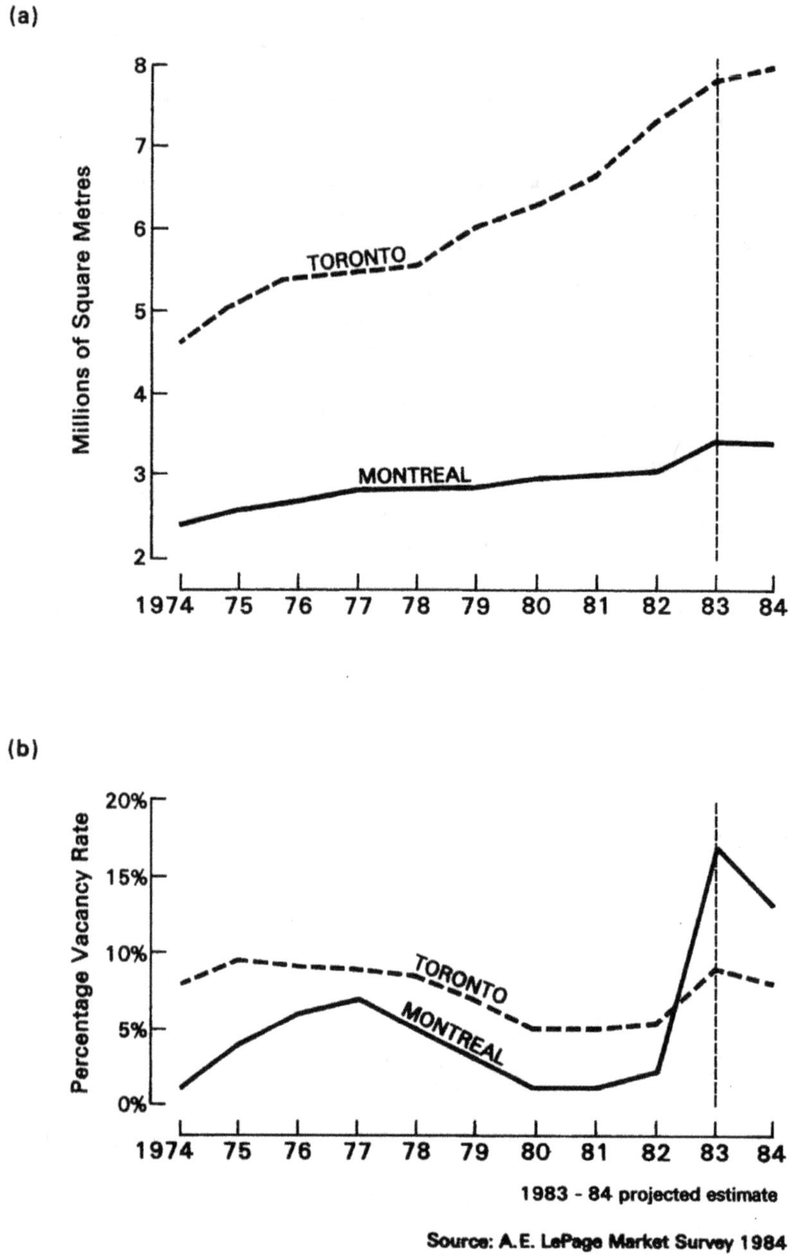

Fig. 7.4 Montreal and Toronto (a) Total office space (b) Percentage vacancy rate

policy on the private office sector and indeed attempted to spread the impact of its withdrawal from the private sector. Nonetheless, this was not always possible, particularly since there was a natural desire to upgrade office space as the opportunities arose, leaving a stock of office space vacant in need of refurbishment before it could be re-let in a weakened market.

By the late 1970s, the policy had begun to have its effect. Significant government offices had been relocated in Hull, including by 1977, the Departments of Consumer and Corporate Affairs, Labour and Environment and by 1979, the Canadian International Development Agency, the Department of the Secretary of State and the Department of Indian and Northern Affairs. Office surveys at that time predicted an alarming rise in vacancy rates by the 1980s. For instance, in 1977 the Royal Trust Real Estate's survey of office space vacancy in Ottawa's central core indicated a rate of 14% in 1978, 25% in 1979 and 35% in 1980. Although these predictions were lowered in a survey two years later, to indicate a rate of 22.0% in 1980, peaking at 25.0% in 1981, the general picture was one of a disastrously imbalanced office market as a result of the federal government programme. Some owners of buildings were undoubtedly affected deeply. For instance, an entire central area block, bounded by Kent, Albert, Lyon and Slater Streets, containing both offices and other land ripe for redevelopment was surrendered by Rockford Developments to the mortgage holders, Prudential Assurance of America in 1979, because of the weak state of the market (Ottawa Citizen, January 28, 1979).

In fact the forecast excess of office space in the downtown core of the city did not materialise since the private office space released back to the market was largely absorbed in a manner quite contrary to that predicted. Indeed by 1983, there was a marked shortage of high grade office space in Ottawa. It should also be noted that the government did not withdraw completely from the private office sector. Indeed in Hull, some of its new offices are rented, whilst in Ottawa itself, even by 1983, projects such as the refurbished Centennial Towers (33,500 square metres) were leased by the Federal government.

The survey of office vacancy in the Ottawa Region carried out by the Commercial and Industrial Development Corporation in November 1983 showed that of a total privately owned office inventory in the central area of some 10 million square metres, 3.89% were vacant, whilst in the Ottawa region as a whole, there was a vacancy rate of only 3.4%. On the other hand, the discrepancies in vacancy rates between different categories of offices is illustrated by the fact that in 1980, Class A office buildings had a vacancy rate of 1.0%, Class B 15.4% and the lowest class, Class C, 17.3%. By 1983, however, the renewed demand for offices in Ottawa had absorbed even this supply of lower class buildings, to produce vacancy rates of

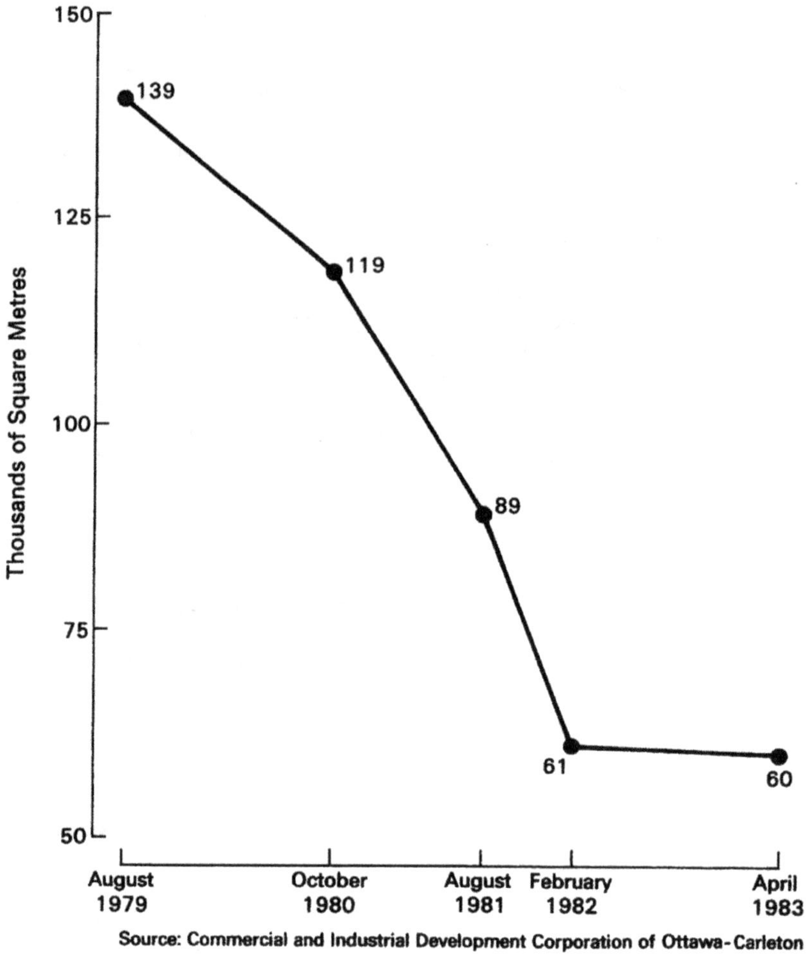

Fig. 7.5 Office Vacancy in Central Ottawa 1979 - 1983

Class B 1.3% and Class C 0.7%. The general trend of declining vacancy is illustrated in Figure 7.5.

Obviously the failure of the office market to match the pessimistic forecasts of the 1970s has to be explained by an increased demand for office space from the private sector. Such demand has been particularly strong from the computing services and other business and professional service organisations. Table 7.1 shows the absorption of office space in the downtown area during 1982 by all organisations.

Table 7.1: Office Absorption in Downtown Ottawa 1982

Classification	No. of leases	Total Sq. metres	% of total
Professional	14	4,667	31.9
Software/Word-Processing	5	2,022	13.8
Financial Services	4	1,068	7.3
Other Services	20	4,846	33.1
Corporations	1	743	5.1
International Companies	3	629	4.3
Insurance	4	446	3.0
Other	2	214	1.5
	53	14,635	100.0

Source: Commercial and Industrial Development Corporation of Ottawa-Carleton

By 1983, the overall effect of the relocation of offices can be gauged by the geographical distribution of all government-owned and leased offices in the National Capital Region, which includes central Hull and peripheral Quebec locations as well as central Ottawa, its suburbs and other Ontario locations This distribution is shown in Table 7.2.

Table 7.2: Distribution of all Federal Government Office Space in National Capital Region, 1983

	Sq. Metres	% Total
Hull Core	553,936	17.2
Other Quebec locations	71,944	2.2
Total Quebec	625,880.7	19.4
Ottawa Core	1,309,937	40.7
Other Ontario locations	1,286,994	39.9
Total Ontario	2,596,931	80.6
	3,221,811	

Source: Public Works Canada.

The FORP still had as its aim a 75%/25% division of federal office space between Ontario and Quebec, but it is clear that in a relatively short period of time the Federal government had brought about a major locational shift in its office-based activities. Rarely have central governments had such a direct impact on the office market of a particular city, but nonetheless, the case of Ottawa illustrates the potential power of the public sector in influencing the office development industry.

NORTH AMERICA - SOME CONCLUDING COMMENTS

This chapter has demonstrated the varied patterns of office development both spatially and temporally. In some cases, the process has been stimulated by economic conditions, such as those which triggered the Houston office boom of the 1970s. In others, governmental policies have affected the process sometimes quite dramatically, as illustrated by the Canadian case discussed above. North America also presents the spectrum of office control measures ranging from the virtual *laissez-faire* of Texas to the relative control of Canadian cities. In all cases, but especially in US cities, it is evident that overseas investment has had a large part to play in the development process.

It is equally apparent that the office market has been highly varied both in space and time. To talk of one development cycle is misleading since different cities have been subjected to varied factors. Equally, North America demonstrates the intra-urban variation in office growth occasioned by its well-developed suburban market. In the case of New York, it could be argued that a series of development cycles have affected the region's office markets, with the city centre and suburban markets being affected at different times.

In terms of overseas investment, the strengthening of the US dollar in 1983-84 might reasonably be expected to have a restraining effect on new investment from overseas, since the cost of buying into the market has risen above that of the late 1970s when the currency was weaker against European currencies. On the other hand, the rent returns on existing US property holdings have increased in value to make them even more valuable assets in property portfolios of the institutions which have already invested funds there.

The following chapter considers new office technology and its likely impact on the office development process. It remains to be seen whether this particular market, in which one would expect to see early use of new technology, will be equally responsive to the new and emerging requirements of the modern office as it has been to the demands of the 1960s and 1970s.

Chapter Eight

OFFICE DEVELOPMENT AND TECHNOLOGICAL CHANGE

In our analysis of office development, it has become clear that the suppliers of the product have a key role in determining its geographical distribution. They are of course influenced in their decisions by questions of the relative profitability of investments, involving both intra- and international comparisons. It is becoming increasingly clear, however, that technological innovation in the sphere of office activities may require a re-appraisal of the continued economic viability of office investment. In turn, this will lead to questions concerning both the activities of the office development industry and the location of future office development. This chapter attempts to sketch the outline of the changes in office technology and to consider their likely impact on the development process.

All office transactions involve the acquisition, handling, storage, processing and communication of information. The submission of a motor insurance claim, the placing of an export order for computer hardware or an application for a loan for industrial development all involve that somewhat intangible commodity - information. Information in all its various guises is subjected to a range of processes, all of which are office-located. Office technology has always striven to improve the process of information handling in a general sense and there can be little doubt that recent advances in so-called information technology have very considerable potential advantages for all concerned with the handling and processing of information.

The improvement in information handling is of course, nothing new. The mechanisation of office functions dates back to the second half of the nineteenth century. The typewriter and the telephone were introduced on a commercial scale in the 1870s, with Remington, for instance, producing typewriters commercially in 1876. By 1900, more than 100,000 typewriters were in use and annual production was exceeding 20,000 per office. (Giuliano, 1982). Cost reductions and improved technology made the typewriter commonplace and universally

DOI: 10.4324/9781003174622-8

accepted. There are close parallels with the development of late twentieth century technology except that the process of change and innovation is now even more rapid than that of a century ago. Thus today's office is either undergoing a large scale transformation to accommodate and employ new technology or at the very least is faced with assessing very carefully the potential of the new technology. Delay in its introduction will be, one suspects, only temporary.

The changes in office practice either brought about or at least heralded by the use of the micro-processor, have a potentially far-reaching geographical impact. The alternative scenarios for future office operations and consequently the office development process vary quite widely and it is inevitable that any attempt at assessing future trends in this field raises at least as many questions as it provides answers Yet it would seem blinkered to ignore current developments particularly because on the one hand their potential impact and magnitude of operation far outweighs that of their nineteenth century counterparts and on the other, any questioning of the direction of the office development process has considerable importance for the office development industry. There are a number of questions which deserve our attention. Firstly, what are the important changes and developments which are already affecting offices? Their full impact has yet been felt by only small proportions of offices but many more are beginning to come to terms with the changes. We can best answer this question by examining closely the field of information technology, which itself has been perhaps something of a "buzz-word" of the 1980s, mirroring the "white-hot" technological revolution of the 1960s. A second question relates to the future geographical patterns of office activities and asks whether there is likely to be a fundamental re-shaping of the locational distribution of offices stemming from the introduction of new technology. Fundamental to this question is the extent to which new technology has been adopted by existing office users, since we are not merely dealing with hypothetical changes, but with real office organisations, subject to social and economic forces which may indeed run counter to large scale changes. Finally we may enquire what effect these changes are likely to have on the office development process in general and its geography in particular.

THE TECHNOLOGICAL CHANGES - THE NATURE OF INFORMATION TECHNOLOGY

Despite the popularity of the term, clear definitions of information technology do not abound. In a general sense we are talking about the automation of certain processes especially in the non-manufacturing sector which has been enabled by the introduction of new machines and devices heavily dependent on the micro-processor. In a study of the potential impact of information technology on the office property market, it has been seen to include "the creation, manipulation, storage, duplication and transfer of different types of information in an electronic format" (CALUS 1983, p.15). The same study points out that the development has been a direct consequence of a new sharing of technology by companies, office machines and communications, made possible by digitisation. In other words, information of all kinds, whether handled by telecommunications, typing and copying devices, or data processing units, can now be handled in a common format of binary digital units. This important technological convergence has brought a common base to office functions, and with it, the real possibility of a fully automated office, in which a unit of information can be received from an external source, manipulated or processed and then displayed or prepared for further onward transmission without the need for its physical transcription or handling at any stage.

Although the precise facilities related to information technology in any one office are impossible to predict, it seems reasonable to accept the National Computer Centre's suggestion that five principal elements will be involved. They are work stations, local area networks, central facilities, a private branch exchange and external facilities. These may be combined in different configurations as Figure 8.1 indicates. The nature of each of the five elements needs to be understood before a judgement concerning its effect on office location can be made. In summary, therefore, each is described below.

Workstations

Workstations are essentially individual but interlinked working areas containing the equipment necessary for an individual to receive or generate information, to enable its processing and its manipulation. Some workstations can be extraordinarily sophisticated, giving some credence to the concept of the "paperless office". For instance, the Xerox 8010 Star electronic desktop can serve as a small computer, a word processor and a generator of graphic material (Giuliano 1982). The desktop itself is displayed on a screen which shows a number of "icons" representing items usually found in an office such as filing drawers, in-trays, out-trays etc. Using a keyboard and a "mouse" (a small hand-held module) to indicate movement of objects, the operator can manipulate the items on a screen, open access to documents and create files to store information. All of this use of the screen is analogous to more normal office activity. It is perfectly conceivable that similar tasks may be performed in the future *via*

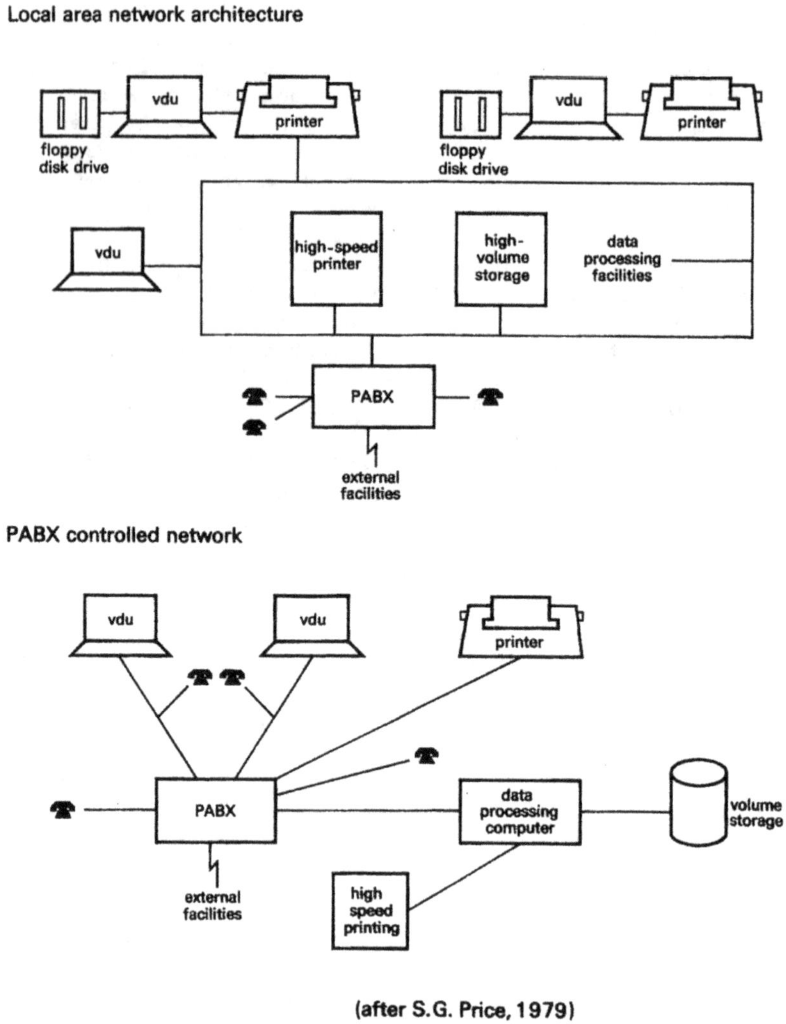

Fig. 8.1 Alternative Arrangements of Electronic Office Facilities

(Reprinted by permission of the National Computing Centre Ltd. from "Introducing the Electronic Office".)

touch sensitive screens on which an operator indicates the required movement of information. Such workstations are likely to be linked together in a so-called local area network (see below).

The potential for workstations to be set up in an office employee's home is obvious, particularly as it could be argued that travel and energy costs have risen as telecommunications have become both more sophisticated and in relative terms less expensive. There may be, however, other powerful social and economic factors which would hinder this process.

Local Area Networks

A Local Area Network (LAN) is essentially a system for linking together computers, workstations, memories and printers so that information can be shared rapidly and effectively between them. Often this would involve small computers and quite possibly, dissimilar information-based systems acquired from a variety of sources. Thus the various pieces of hardware within a company's office are able to communicate one with another. Once this is established, it is perfectly feasible to institute an electronic mailing system by which internal communication can be made without recourse to paper, using the medium of visual display units (VDUs) at individual worker's work stations. Any printer can be operated by any terminal and all data memories are accessible. The potential operating savings for an office with such an all-encompassing system are very considerable. LANs are an advance on ordinary telecommunication links in that they offer highspeed transmission with low levels of error. The methods by which this is achieved may vary but essentially involve multi-accessibility to a data "highway" along which data "packets" are transmitted, bound for specific destinations. The Xerox company have developed a system called CSMA/CD - meaning carrier sensitive multiple access with collision detection - to ensure that each data "packet" is carried through the system from origin to destination without being confused with other packets of data. This depends on the sending stations sensing when the system can accept a data packet without the risk of its colliding with others in the system. This system uses a single coaxial cable and is termed a baseband technique in contrast to other techniques which are broadband, utilising a coaxial cable capable of carrying a number of separate signals (including video signals, since the same technology is at the heart of cable television systems). Current trends in LAN technology appear to favour the more widespread introduction of the latter technique.

Obviously LANs offer tremendous potential for communications within the office. However, the notion of remote office employees conjured up by the workstation described above, recedes since the systems are essentially local in character. Buildings in reasonably close proximity may be linked together in an LAN, but the linking of a suburban workstation and a downtown central office would be very expensive indeed. However, the introduction of LANs has very important implications for office development since the requirements of office design may be very stringent to ensure the neat and cost-effective installation of a LAN. It is ironic that some older offices with high ceilings and the capability of housing new and possibly complex wiring systems, may have an advantage over new offices of more restricted height into which it would be difficult to introduce false ceilings or floors to carry wiring networks.

Central Facilities

This broad term covers a wide variety of electronic hardware whose functions include the storage, filing and retrieval of data, its processing and printing. The actual requirements of any one office will hardly ever be coincident with those of any other, simply because the scale of operation and specific tasks to be performed will vary greatly. Nonetheless, it can be argued that most companies of any size will require a central processing unit, although its scale may be fairly small. On the other hand, major innovations such as IBM's Audio Distribution System, enabling the storage and onward transmission of the human voice following its conversion to binary digits, are likely to be found only in very large scale organisations.

It is worth noting that early word processors generally worked on a shared logic system, comprising a number of screens and keyboards linked to a single central processor and single print station. More recent developments have moved, however, towards shared resource systems, within which a number of workstations each have their own powerful microprocessors to carry out tasks including the running of word processor programmes. These are linked together *via* LANs to a centrally housed electronic high capacity memory and printer, both of which are high cost capital items. It has been argued that an effective LAN in a large organisation may well be dependent on an effective indexing system for information, filing and retrieval with very large data storage requirements. This would call, therefore, for a powerful mainframe computer facility. Manufacturers at the moment, offer systems which can stand alone or be linked to host mainframes according to the needs of individual customers. For example the IBM 8100 Information System and its Display Writer word-processor system may stand alone or be linked to a host mainframe.

Once again, the prospect of major locational dispersion of individual office functions seems an unlikely propsect once it is realised how far individual officeworkers are tied electronically to centralised systems. It is also inescapable that the office space requirements per user have risen and are rising inexorably as centralised facilities require generous space provision, often of a very specialised nature within the curtailage of office buildings.

External Facilities

Access to external facilities involves the establishment of a central information store to which individual users are linked. They can then retrieve specific information from the datastore. Such systems have been developed in the last decade in a number of countries. The general term for them is "videotex" systems, although in the UK they are more commonly known as "Viewdata" systems. A full service has been offered in the UK since 1979, with the Prestel system, whilst internationally, the other leading developers of videotex systems have been France and Canada. The advantages of a viewdata system mean that an individual untrained in data processing techniques can access the system using a suitably adapted television receiver rather than a computer terminal. The actual link is provided by an ordinary switched telephone circuit rather than a coaxial cable or other specially installed link. The system can be developed in order to permit access to a number of computerised data bases linked to the system and furthermore, to accommodate two-way transactions. This opens up the possibility of electronic banking, booking of travel tickets and many more such transactions, all of which are now beginning to be introduced in the UK. The addition of the "Gateway" system to Prestel in 1982, permitted these developments of the British system, but similar advances are being made elsewhere to ensure that such external systems are becoming something far more sophisticated than merely information retrieval systems.

It is interesting to note that in Britain, the marketing of the Prestel system was originally aimed at both the domestic and the business markets. Sales to the former were, however, very low, amounting in 1982 to no more than a few thousand in the entire country. The business community on the other hand have been more receptive, with the travel business being the sector which has made most use of the system, followed by those concerned with financial, stock and commodity markets.

Besides the publicly available viewdata system, several companies have established their own private systems for their own purposes. The companies and organisations include such varied concerns as Ford UK, Allied Breweries and the Lothian Regional Health Board, but it is noticeable that the travel trade are well represented with Thomas Cook, Thompson Holidays and Horizon each having its own private system, whilst those in the finance sector with private systems include American

Express, Barclays Bank, the Co-operative Bank, the Trustee Savings Bank and the Stock Exchange. (Financial Times, 1 December 1981).

The technical development of external services has been very rapid, but take-up of the new technology by office-users was rather slow. Nonetheless, recent developments such as the addition of "gateways" to viewdata systems offer very considerable potential to the office user, offering not only a rapid access to information, but the facility to process certain business transactions very rapidly. The systems are therefore of benefit to the business user, although it is less clear that they will have any strong locational effect on the office itself.

THE REALITY OF THE NEW TECHNOLOGY

The technical feasibility of introducing to almost any office much of the interrelated "hardware" of information technology is not open to question. This in itself however, does not ensure that the conventional office is doomed to a rapid extinction. As the CALUS report on the impact of information technology observes in summarising some of its conclusions,

> Visions of the future based solely on technological feasibility are inaccurate: the technology must be placed in a wider economic, social and cultural context. The availability of the technology at an affordable price is a necessary but not a sufficient pre-requisite for the introduction of IT. (CALUS 1983, p.14).

The attitudes of office employees and their unions has already proved to be a critical factor determining the pace at which new technology is introduced. The state of the economy is one factor rarely cited, yet an expanding economy is more likely to encourage a company to seek the cost savings of IT as an alternative to increased labour costs, than one in recession. Both the type of office function and the size of the office unit are also important variables to be considered. Certainly the small solicitors' practice in a market town is less likely to be an innovator in this field than a major insurance company or building society. A certain amount of research has been carried out to investigate which firms are likely to be the users of new technology. The early work in this field in the 1970s concentrated on innovations in telecommunications (e.g. Pye 1979). The Communications Study Group at University College, London has examined the potential role of new telecommunication techniques. In controlled experiments, for example, it was shown that for certain kinds of office activity, different modes of communication gave differing results. Thus it was found that it was more likely that the opinions of others could be changed using audio-only communication compared to either audio-visual or face-to-face contacts. In general however, users themselves did not like audio-visual or audio-only communications for interpersonal contact. The whole thrust of this research and that of others examining the potential of new services such as teleconferencing (e.g. Pye and Williams 1977, Short 1970) was to demonstrate their impact on possible office decentralisation particularly from a congested central location such as London. In that sense much of this empirical material is of limited value to the present problem, although we might note in passing that the innovation of teleconferencing or confra-vision have found only limited acceptability and are not universally available. Of more direct relevance is the research by P. W. Daniels involving the analysis of over 300 offices in provincial towns and cities in the UK. Although the firms examined were all based in the provinces rather than in London, the conclusions reached are sufficiently important to warrant our close attention (Daniels 1983).

The project involved a survey in 1980 of 304 business service offices in eight provincial centres. Each office was investigated to determine the degree to which there was an awareness or use of each of nine information processing and telecommunications (IPT) items. They varied from the fairly commonplace, such as internal telephone systems to more sophisticated equipment such as word processors and computers. Certain items were obviously well known, but of others such as audio-visual and data transmission facilities, there was relatively little knowledge. For instance whilst 82.3% of respondents possessed internal telephone systems and only 1.8% claimed no knowledge of such systems, only 10% had Datel data transmission facilities, with 52.2% claiming no knowledge of them. The study was able to suggest that the size of the office was an important factor in determining the extent to which the items were utilised. Obviously larger offices were more likely to be using the more sophisticated equipment such as word processors, telex systems and externally based computers. Offices which were branch offices of parent companies elsewhere would similarly have a need for certain of the advanced communication equipment. Daniels' study examined office functions, making the distinction between the insurance, banking and finance group of functions and professional and scientific services, to determine whether this was an important variable determining the use of new technology. His findings indicated that there was little difference between the two groups in their use of the new equipment, although banking, when considered separately did have a higher level of use than was expected statistically of data transmission and computer services. These findings notwithstanding, it would still seem reasonable to suggest that certain office functions are more likely to use some items of new technology than others. Thus banking and insurance branch offices may have a very high requirement for data transmission facilities whilst a consulting engineers will require other facilities such as on-site computing facilities with graphics capabilities.

The study notes that the possession of one item of IPT by an office increases the likelihood of the use of other related equipment. Certainly as Daniels notes (Daniels 1983, p.29) this is partly because of the inter-dependence of some kinds of IPT. One could add that there is an increasingly greater potential for interdependence of office equipment as the technological convergence referred to above increases the compatibility of different kinds of office equipment. The provincial offices study demonstrated a "triangle of interdependent facilities", linking word processors, telex and internal computing systems. Daniels concludes on this point that "the 'pivotal' IPT facilities in provincial offices, therefore, appear to be telex and internal computer facilities, both of which will increase the chance that the firms with these facilities will also choose to utilise other equipment." (Daniels 1983, p.32).

The study also included an assessment of attitudes to new technology. Certain items such as computers gained high acceptance, whilst others, such as word processors, were treated much more cautiously, in part no doubt due to the early and therefore uncertain stage of their technical development. The overall classification of the surveyed firms by Daniels is an important pointer to the degree to which the office development process itself will be affected by the demand for new office facilities. Three classes of firms emerged, comprising a) a minority of positive adopters, b) a much larger group of conservative adopters and c) a residual group of traditionalists. (Daniels 1983, p.39). The essential point is that whereas office equipment manufacturers may erect elegant scenarios of the paperless office of the future, that will emerge only if the office workers themselves, at all levels, accept both the economic and social *raison d'être* for large-scale change.

Whilst empirical work such as that by Daniels could be said to sound a cautionary note to counterbalance the claims of the office equipment industry, it is nonetheless true that those concerned with office development do need to look very carefully at the implications of the new technology both for future schemes and for existing buildings.

J. Hennebury has suggested that the locational impact of the adoption of new office technology may well be determined by the pace of innovation (Hennebury 1982). In the event of there being a very rapid introduction of new technology, it is highly unlikely to have a major locational impact. The reason is simply that it is just not possible to renew anything more than a very small percentage of office space in any one year. Indeed the reality is that the existing locational investment in offices represents a very considerable inertia element built into the system. On the other hand, a more gradual introduction of new technology may bring about a greater locational change as developers have time to adjust to new locational trends arising from the needs and opportunities of the new technology. Thus, the rapid introduction of micro-electronics into offices is likely to lead to pressures to adapt existing buildings. This in turn, may have considerable implications for investment in office refurbishment schemes which have already begun to attract institutional investment.

The adoption of new technology is seen by M. Digby as a further encouragement to decentralisation from major cities, including London. (Digby 1982). He sees the centre occupied by top management and essential business services, with other support staff in decentralised locations. Such locations would be suburban greenfield locations. Certainly this view accords with the concept of the office park, commonplace in North America and becoming more accepted in Western Europe.

The possibility of a more complete deconcentration, taking the form of home-based office workstations in which office employees carry out tasks remotely, utilising tele-communications links to a central base, is an element in a number of possible views of the future. Yet despite the available technology, there is little evidence of such a trend except for certain types of work. Portable computer terminals for sales staff are certainly in use and some companies are experimenting in other areas. For instance, Debenhams PLC, the UK department store group have set up a private viewdata system for directors to receive in their own home at weekends, details of the previous weeks' sales throughout the store group. Using a linked computer, such data can be further analysed (reported in Financial Times, 19 March 1982).

From the point of view of the demand for office space, there is little evidence that any such decentralising trends would in any event bring about a significant reduction. A greater amount of space is certainly required for the new equipment and in general the space required per office employee has been steadily rising in the recent past.

The evidence thus far points to certain future trends for office development. One of the most important is that existing offices will need to be adapted to accommodate the new technology. Even those of relatively recent origin were not designed to take the amount of wiring required by all the artefacts of modern information technology. Thus buildings will have to be modified to take supported (or platform) floors, as well as suspended ceilings, and at the same time, heat and noise insulation will need to be improved to meet rising environmental standards.

A second trend is for office buildings themselves to be more flexibly designed to accommodate change. Certainly institutional investors will not want to be associated with office projects which are in danger of being rendered obsolete by a rapidly advancing technology. A report on the requirements of buildings for the needs of information technology published in 1983 pointed out that buildings with low floor to ceiling heights, services embedded in the building fabric and air conditioning systems which were incapable of adaptation, were particularly susceptible to premature obsolescence. (Strohm 1983). When the economics of adaptation show that costs range from £250 to £1,110 per square metre, and average construction costs of new buildings is approximately £330 per square metre, one can see the clear possibility of certain buildings being rendered useless as a secure investment for the future.

A commonly foreseen trend which new technology is encouraging is the growth of small office units. Fragmentation of the office function would certainly encourage this reversal of the concentration which has been more usual in the past. If indeed we witness a fall in the number of office workers, this could lead to a greater sharing of common facilities such as reception and recreation facilities by office organisations under one roof.

Finally, and perhaps most importantly from the geographical point of view, there is a considerable measure of agreement on the fact that the acceptance and introduction of information technology is leading, and will continue to lead, to greater locational choice for the office user and hence for the investor in offices. The CALUS report, cited previously, sees the complementary pattern of a core of existing office centres, with good communication facilities accompanied by an established network of support services and a series of environmentally attractive decentralised locations. The purveyors of the new technology themselves have already shown the way. IBM (UK)'s headquarters building at North Harbour, Portsmouth, built on land reclaimed from the harbour, enjoying very high levels of environmental design and remote from any other office centre is but one example. Further investment in offices may well be emphasising the needs of versatility or flexibility of design, looking for sites of high environmental quality rather than the requirements of a high contact environment in the city centre. There is no doubt that office activity is changing in a process not dissimilar to that by which industry was transformed by the Industrial Revolution. The rate of change is potentially very rapid, although there are a number of built-in checks on the process, both societal and economic. The inertia induced by the accumulated investment in "conventional offices" is also a powerful conservative element in any tendency towards large-scale locational change.

IMPLICATIONS FOR THE OFFICE DEVELOPMENT INDUSTRY

There can be little doubt that those concerned with office development will become increasingly aware of the need to reassess their investment in terms of its ability to house the office functions of the future. The locational inertia of office development may in itself prevent the widespread dispersion of offices away from established locations. On the other hand, existing office buildings will be carefully assessed to determine whether they can accommodate modern office facilities. Those which are not able to do this are likely to have a relatively short life and be replaced by offices which are designed for this new technology, with a consequently more secure investment value.

Chapter Nine

POLICY IMPLICATIONS

The studies of the office development process, operating at a variety of scales have raised a number of important questions concerning the process of office development and its geographical impact. Many of them relate to the distributional effect of the current process, set as it is within the framework of the system of finance capital of western societies. As it operates at the present time, in most countries, it is remarkably free of controls, and those which are imposed are frequently either ineffective, or merely temporary constraints on the system. Should this process be subject to a greater degree of control and if so, what type of control mechanisms should be imposed? Should the control be on financier, developer or occupier? What have the experiences drawn from the rapid transformation of successive cities' office stocks to teach policy-makers in the next decade and beyond? The pace of development, despite short term effects of recession, has in overview been dramatic over the past two decades and there is little evidence to suggest that the process will prove to be any less pervading in the future. If that is the case, should the office development industry be permitted to impose its requirements on the economy, morphology and social life of cities without requiring a measure of public accountability? As we saw in Chapter Four, community considerations are beginning to question the hitherto inevitable results stemming from a development process in which profit for investors has been the primary, and some would argue, the only consideration. How far is it possible to follow this particular path to formulate public policies which provide a beneficial outcome to all parties concerned? Technological change is of course not new in having an impact on offices, since the telephone and typewriter revolutionised communications last century. The pace of change, however, is more rapid now than ever before. What are the real implications for policy-makers of the technological advances which are affecting communications and the operation of office processes generally? We cannot pretend to have answers to all of these questions and indeed the answers are rarely simple. Yet the issues deserve further

DOI: 10.4324/9781003174622-9

exploration which is the aim of this final chapter, in which the subsequent implications of the office development process for policy-makers are discussed.

CONTROL OF THE OFFICE DEVELOPMENT PROCESS

On the face of it, offices as urban and regional economic functions should be perfectly amenable to control. Since the operation of an office is not tied to a locationally constrained raw material and only rarely are its costs of operation determined by transport costs, one may argue that its location is relatively 'foot-loose'. Advances in telecommunications have undoubtedly assisted in loosening the locational ties of office activities, although research has shown that offices are some way from total locational freedom in terms of total operating costs. (See for instance Pye 1977). Nonetheless, compared to manufacturing functions, office activities enjoy a locational freedom which should permit a greater degree of control and public policy direction.

There is of course no consensus on the need for control. The established property press, much used in this book as a source of information, sees any form of control as a threat to their activities and an unnecessary intrusion in the development process. Such views are nonetheless short-sighted and ignore the powerful social and economic arguments in forms of control and direction. Office activities bring employment opportunities and choice to a community and, as Yannopoulos has shown, may frequently be important employment multipliers in a local economy. (Yannopoulos, 1973). To permit such activities to be concentrated in certain areas to the point of congestion, whilst other areas have a weakly developed office sector is damaging both to those communities with offices and those lacking in these activities. On the other hand, the encroachment of offices into non-commercial areas of cities, transforming local economies and destroying local communities serves only the interests of the development industry. Nonetheless, even if the concept of control were to have general acceptance, important questions remain relating to the timing of control, its precise nature and that part of the development process which is most amenable to control.

THE NEED FOR AN AWARENESS OF THE PROCESS

Before any controls can be proposed and implemented however, there are certain pre-requisites to be satisfied, amongst which is the overwhelming need for an awareness of the operation of the development process by policy-makers, including both politicians and planners. Too often in the past, the full implication of a major legislative measure has been seen only in retrospect. A notorious example was the immediate effect of the embargo on further office building in London imposed in November 1964, to become known as the 'Brown' ban after the Labour cabinet minister most concerned with the decision. With it came the introduction of Office Development Permits and the natural conclusion would have been that a control structure had at last been enforced on the rampant speculation in office property which characterised London at that time during its first post-war office boom. Two major side-effects were, however, equally important at least in the initial period following the enactment of the new legislation. On the one hand, as Alexander states, "on hearing of this move shocked developers who up to that time had been constructing offices in Central London with more than their usual avarice, rushed to sign as many contracts as they possibly could before the mid-night deadline (of 4th November)". (Alexander, 1979, p.63). Thus the immediate impact of the embargo was not to stop development but, in the short term, to encourage it. On the other hand, the effect of the legislation was to push rents upwards in London since an artificial shortage was created. Perversely, a rising market would enable certain developers to hold property vacant with the assurance of securing a lease with an enhanced rent at a later date.

A recognition of the likely impact of such measures amounts to the need for an awareness of such "trigger" mechanisms as they were described in the opening chapter of this book. The decision of the British Conservative government in 1979, to lift exchange controls opened the way for a massive outflow of funds into overseas investment, which it is difficult to believe was entirely foreseen when the action was taken.

The net outflow of investment capital in all activities in 1981 had risen to £5,103.6 million, compared to £2,709.7 million in 1978 prior to the lifting of exchange controls, and 1980-81 net investment in property overseas amounted to £541.5 million, compared to £15 million in 1978-79. (HMSO 1983). Certainly such large scale changes in the geographical flow of investment capital, including that into property, needed to be properly anticipated and their implications considered.

One of the major difficulties relating to the awareness, or otherwise, of the operation of the development process lies in the previous nature of control policies. In the UK., regional policies have been typified by a succession of short term measures, of which encouragement for the tertiary sector was a somewhat belated one. The changes in geographical area in which office development permits operated were frequent and hardly gave sufficient time for a proper and thorough evaluation of their real impact. The French emphasis on their *métropoles d'équilibre* in the middle and late 1960s to establish tertiary growth poles in the provinces was quite rapidly followed by a policy favouring the *villes moyennes* or medium-sized towns, and then by policies concentrating on small towns and their development. (see Scargill 1983). Switches of policy such as these prevent a proper evaluation of their economic effects although politicians are always able to measure their political effects in a remarkably short period of time. It is this dimension of policy-making which has encouraged short-term solutions since in countries such as the UK and France, regional policies and indeed any other attempt at redistributive mechanisms are subject to the vagaries of political control. Only rarely have countries in the West been in a position to pursue particular aims of economic planning for protracted periods without the need to weigh political considerations. One such example would be Sweden for most of the post-war period where clearly defined national objectives have been pursued without wild swings of political doctrine.

This brings us to a further need for awareness by policy-makers in formulating controls on the office development system since all such controls are only really acceptable and viable in the framework of a nationally accepted plan. Ambrose and Colenutt, whose conclusions are further reconsidered later in this chapter, called for a National Building Plan to identify development needs in different parts of the country and to estimate "the number of housing units, factories, shops, offices and community facilities that are required". (Ambrose and Colenutt 1975, p.165). Whilst such a plan may be thought to be excessively optimistic in terms of relating economic and social needs to specific building requirements, there is a powerful logic in calling for an overall strategy within which planning policies and control measures can be properly formulated and implemented. Outside the world's planned economies, such foresight is rare, although we may point to France as an exception. Since 1947, a succession of National Plans have been formulated by the French government to guide the development of the economy in general. Nevertheless, it must be pointed out that outside economic influences made it increasingly difficult to adhere to specific objectives, and led to the abandonment of the VIIIth Plan for the period 1981-85. (See Tuppen 1983, p.21). The argument remains, however, that without an awareness of the general needs of an economy, both economic and social, it is difficult to formulate consistent development control policies at either a regional or an urban level. In Britain the introduction of structure plans since the 1968 Town and Country Planning Act has gone some way to defining sub-regional goals, within which policies at that level can be formulated. Such policies relate both to economic and social goals and are important determinants of local plans within which the development control system can be operated. Above the structure plan level, however, no such consistent strategies have been formulated. Instead, each government has pursued its own broad policies motivated by its own particular philosophy.

AWARENESS OF NEEDS OF OFFICE USERS

If an awareness of the operation of the system and of the framework within which it is intended to operate, is an essential pre-requisite to defensible and realistic policies for office development, no less important is the need to be aware of the requirements of office users. In a sense these are not too far removed from the requirements of the office development industry since only those offices in demand will be seen favourably by the latter. On the other hand, some categories of office user may have requirements which are not normally considered by developers and investors. Indeed those offices most favoured by the latter may be suited only to certain types of office users such as major tenants occupying large areas of floor space. An additional major requirement is for policy-makers to be aware of the needs occasioned by the new technologies discussed in the previous chapter. Rarely, and perhaps only in certain areas of retailing, has it been necessary to categorise as obsolete, buildings of scarcely twenty years' standing. Yet in the case of offices, an early obsolesence is by no means uncommon. A complicating factor is that many buildings of the 1950s and 1960s - the first property boom in the UK - were built prior to the more stringent demands of energy conservation. The successive generations of office building at La Défense in Paris, considered in Chapter Five, demonstrated the changing requirements of office occupiers over a short period of time in this respect.

It is not only energy considerations, however, that have altered offices since the layout requirements of offices have also altered with a move towards open-plan design, superseding multi-storey offices with each floor sub-divided into separate offices. Nonetheless the major factor leading to premature obsolesence is the increasing demands made by new technology. Costs of adapting offices to house new equipment may be prohibitive, a factor which is particularly important in the early post-war office buildings using space-saving building techniques with shallow floor and ceiling spaces. The real impact of these new requirements remains to be evaluated, but the signs thus far are not promising. The situation is exacerbated by the fact that in a market in which there is a healthy supply of new space available, a general tendency emerges to improve office accommodation leading to the vacating of older office blocks in favour of those better equipped. For many office users, demands of prestige and company image may be as important as the building itself, but whatever the motivation, the result is a growing supply of relatively poorly equipped office space which it maybe difficult to refurbish to cater for current technological developments. The experience of the Ottawa market under the rather artificial conditions discussed in Chapter Seven, demonstrated the tendency for an up-grading of office space by tenants to take place. Whilst in the past, a degree of refurbishment of the vacated offices might have equipped them for the market, it may not be quite so easy in the future if a broader range of potential tenants is demanding office space capable of housing the new office technology dependent on LANs and their circuitry requirements (see Chapter Eight). There is then every justification for policy-makers making themselves aware of these new requirements and ceasing to regard all office stocks as undifferentiated floorspace for office activities. The difference between the general store and the hypermarket in retailing terms may be even more exaggerated in the comparison between the office of the 1960s and that needed for the 1990s.

A second awareness of the needs of office users arises partly from the foregoing discussion of new office technologies. At the present time, the development industry concentrates its activities on securing a limited number, and often only one, occupier for a new office block. In the terms of the property profession, it is often seen as a relative failure to have to sub-divide an office building to let to a number of tenants. Financial institutions of course exert a certain pressure in this regard since most investors prefer to see one tenant of good standing secured in their building rather than an agglomeration of small concerns each with differing lease provisions and conditions. Only as a second resort therefore will a developer accept the necessity to let a building in smaller units. Until recently, this practice has been reasonably acceptable and the arguments in favour of it from the property industry's point of view, understandable. Indeed it has had its advantages since as major companies have moved to new buildings, often consolidating their activities in the process, they have frequently vacated a reservoir of smaller offices which are available for office users with smaller space requirements. The introduction of new office technology, however, may mean that there is a need to provide new well serviced space for smaller office users as a matter of course, in a manner not unlike starter units for small industrial premises. In the latter case, suitable workshop facilities disappeared for small industrial firms so the requirement for new purpose-built accommodation increased. In the same way, if older premises prove to be too obsolete for even small firms and organisations, so there may be a need to be aware of these requirements and formulate planning and development policies accordingly.

THE NEED FOR AWARENESS OF THE OFFICE DEVELOPMENT INDUSTRY

It is not possible to formulate policies concerning office development to control any sector of the industry without a thorough knowledge of the industry itself. At the present time, our knowledge of the office sector of the economy is generally limited to the numbers engaged in the activity and their location, derived from *inter alia*, census information. Specific studies by local authorities may add to this, but our only knowledge of the complexities of the industry itself is based on piecemeal information gleaned from a variety of sources. The industry itself does little to assist public authorities in this respect, since the two are most often seen as opponents in a planning conflict arena. Schwartz, in discussing New York real estate, talks of "inadequate information... conditioned by the fundamental secrecy of the real estate sector." (Schwartz, 1979, p.233). Such secrecy is apparent in many aspects of the property sector and does not assist policy-makers in assembling a comprehensive picture of the office development industry. On the other hand, the commercial requirements of the industry may often demand secrecy, particularly in respect to rent levels "achieved" on new lettings of office property and similar transactions. The fact remains, however, that an enhanced awareness on the part of individuals, most of whom are indirect owners of office property in what would be to them quite exotic locations, of the property transactions of insurance companies and pension funds, might in time promote a political will leading to a more open control of the development industry. At the present time one is left with the impression that a limited number of individuals, wielding considerable financial weight, are able to make far-reaching locational decisions having great social and economic significance, with little public or political awareness either of the real concentration of such power or of the way in which the system operates. Only recently has there been evidence of a questioning of such lack of social accountability, illustrated by the attempts, albeit unsuccessful, by the National Union of Mineworkers to question the property investment policy of the National Coal Board Pension Fund by resorting in 1984 to the High Court, in an effort to place restrictions on the Fund's investment policy.

PUBLIC SECTOR AND PRIVATE SECTOR CONFLICT

The history of property development is well marked with case histories of the difficulties of reconciling the interests of the community and those of the private sector. In the context of property development, the pension funds must be considered as a part of the private sector, despite their manifest public ownership, since they regard their obligations towards their fund beneficiaries in much the same way an insurance companies view theirs towards policy-holders and shareholders. The private sector is defined then as the institutional investors and property developers, for whom development profit, either long or short term depending on the group involved, is the primary reason for being involved in property development. On the other hand, the public sector, comprising national and local government as well as the more general community at large, has other considerations. These are not easy to generalise since the political complexion of government will determine its primary considerations. It can be argued, however, that considerations such as community social benefit as well as economic efficiency should be powerful motivating considerations for the public sector although the balance between the two might be disputed. The concept of planning gains, or the provision of some facility for the public good in exchange for the right to development profit, would be seen by some as the ideal balancing mechanism to reconcile public and private interests whilst to others it is an unworkable compromise. Indeed, as many inner London boroughs have chosen to state in their planning policies (see Chapter Five), planning gains may not be sufficient compensation for the large scale social and economic disruption inherent in the pursuit of development profit by the office development industry.

Studies in this book have illustrated the difficulties in reconciling the requirements and demands of both sectors. Public involvement in the development process and indeed its eventual dependence on it was illustrated by Hong Kong, discussed in Chapter Four. Whilst in the short term, it was possible for the Hong Kong government to act as a land broker for the development industry, it was faced eventually with a situation where its financial dependence on the private sector's unrealistic valuation of land became too great. At a smaller scale, the very close involvement of office developer with local authority can prove to become an area of conflict of interest as witnessed by Cardiff's difficulties in redeveloping the centre of the city in partnership with a major property developer.

The relationship between public and private sectors is therefore delicate and one which must be clearly appreciated by policy-makers. Some would argue that since the interests of the two sectors are irreconcilable, the private property sector should be taken into public ownership, a proposal which is given further consideration later in this chapter. At this stage, however, the potential for conflict and the need for policy-makers to be aware of it needs to be stressed.

CONTROL MEASURES AND THE DEVELOPMENT CYCLE

At the beginning of this chapter, the general arguments in favour of control were advanced - to avoid congestion and spread economic growth opportunity and employment choice. There are, however, other considerations to discuss prior to considering the actual form of controls which might prove to be most effective. One justification for the imposition of controls is that they would prevent the boom and slump cycle of office development which has typified our post-1945 experience. Instead, a control system could regulate supply at times of over-simply and act as a stimulant at times of under-supply. Leaving aside the difficulties implicit in this notion, arising from the long time-lags between identifying need and bringing office space to the market, there are other arguments which have been advanced in at least partial justification for periodic surpluses of offices. They are well put by Schwartz when he suggests that a surplus of offices tends to act as a constraint on central city rents, which, at least in the US context, makes them more competitive with suburban locations, thus restraining the tax-eroding suburbanisation of offices. Secondly he points to the increased ability for office users to up-grade office accommodation in conditions of office surplus. Finally, a surplus may induce favourable lease conditions for an office tenant, "thus tying them to the city for a longer period of time than they would be able to commit for in a seller's market." (Schwartz 1979, p.231). He goes on to cite the initiative of the New York City Planning Commission in 1973 by which the public was asked whether municipal government should intervene to smooth out the extremes of the construction cycle. The results of this initiative was a degree of support from the private sector for intervention during the "downside" of the cycle in the form of tax concessions and other measures to lower rents for occupiers, whilst sustaining development profit. Intervention during the reverse phase of the cycle was not, however, supported. As Schwartz admits, such results were hardly surprising, given the private domination of the property market. On the other hand, public opinion in Britain has often been more in favour of intervention to prevent the oversupply which turns the market into a buyer's market with some of the possible advantages for tenants outlined by Schwartz above.

In practice, in Britain, controls in relation to the development cycle have usually been poorly timed, a tendency strongly criticised by Barras (1984). In reviewing the general problems of controlling the volatile conditions of the London office market, he addresses the general question as to whether development control policy should be used to even out the booms and slumps of the development cycle. One advantage of such a policy would be a reduction in the volatility and uncertainty of the market, which would bring benefits to developers, investors and occupiers. At the same time, the negative externality effects of cleared sites and excessive vacancy of new buildings would be moderated, (Barras 1984, p.46). Despite the reservations of the private sector, therefore, cited by Schwartz, there are advantages to be gained from such intervention. However, as Barras crucially comments, "if this is to be an objective of planning policy, then development control must be exercised in a manner opposite to that so far pursued in the post-war period, when it has tended to reinforce rather than smooth out the cycle." (Barras 1984, p.46).

The relaxation of controls during each upturn in the cycle has been followed by re-imposition during the subsequent downturn of the cycle. In fact, the opposite pattern is required to control the cyclical swings of office development. Labour government controls, in the mid-1960s with the introduction of ODPs and again in the mid-1970s, with the adoption of stronger ODP policies, were imposed after the surplus had been created. Conversely Conservative government relaxation of controls, including the abolition of ODP controls in 1979 have taken place when development cycle upturns have become established. Barras contends that to make matters worse, these changes in development control policy have actually boosted the average level of development profit, simply because the effects of each slump have been decreased by control measures. Timing of controls are therefore of crucial importance and experience in Britain points to the need for a greater consistency of policy and not one based on *ad hoc* requirements of the differing political persuasions of successive governments.

An additional problem which has to be overcome if booms and slumps are to be controlled stems from the system of giving planning consents, which need not be acted upon immediately. In times of low demand, therefore, developers and others can acquire planning consent for an office development and retain it until more beneficial market conditions prevail. This particular problem can become a barrier to effective control and was identified as such in an analysis of the City of Westminster's office development by one of its planning officers. He commented in 1981 that "far more floorspace is being permitted than is being completed, and the backlog of outstanding permissions has grown dramatically. One of the ironies of planning 'control' is that planners are powerless to control when and even whether, this backlog will be implemented." (Stevenson 1981, p.19). Plainly effective control of the development process requires planning consents to be given for more limited periods than the five year life of current consents.

POLICY OPTIONS

The question remains as to which part of the office development industry is most amenable to control at the same time as providing the most effective results of such control. In broad terms, the options are: a) to control the investing institutions, b) to control the developers of offices, especially in terms of the location of development and c) to control the occupiers of office premises. Each has its advantages, although in practice the most common form of control, as illustrated in Chapter Three, has been to control the developer through some form of development control policy. This policy can be imposed at a national level (eg. ODP policy), sub-regional level (structure plan policies) or intra-urban level (development control via planning permissions). Each option is however worthy of some further consideration.

The control of the investment made by major institutions is of course a politically charged proposal. The notion of directing a proportion of funds into specific areas such as industry or new technology has been proposed by many, including Britain's TUC[(1)]. A degree of control over pension funds has been exerted in Scandinavia, some European countries and Australia, yet it raises many questions. The trustees of pension funds perceive their duty as the maximising of returns on investments. In the event of funds being directed to invest in certain areas by central government, there is the possibility that governments should be asked to underwrite losses to the funds arising from such investments, and possibly compensation for lost investment opportunity elsewhere. At a more general level, however, control of institutional investors' activities could be gained by the re-imposition of controls on the outward flow of investment. It is quite another matter, however, to expect the money thus saved from being invested in Denver or Singapore, to be invested in Darlington or Sunderland. Given the requirements of the investors, it will be more likely to be invested in agricultural land in Southern Britain, or an office block in the City. In other words, although the social and economic arguments are strongly against investments overseas whilst the now traditionally disadvantaged parts of Britain require investment capital, it is idealistic to suggest that a switch of investment by any central intervention short of central control can be easily achieved.

The second possibility is to exert control via the developer in some form of development control policy exerted at the level appropriate to the particular problem. Thus the ODP system in Britain represented an attempt to control the regional allocation of offices, although as we have seen, it was largely unsuccessful in that regard. It is worth returning to our analysis of the French office market (see Chapter Six), to examine this particular policy option in more detail. Since the mid-1960s, the stated policy in France has been to stimulate the service sector outside Paris. Within the Paris region, the aim has been to correct locational imbalances, particularly between the east and west of the region, whilst the new towns have since their establishment been singled out for special treatment, including a 20% share of office development permits since 1974. In essence, therefore, a framework has existed at a number of levels to control development. An analysis of office construction, however, suggests that despite the policies, only limited success has been achieved. For instance, a recent study of office development in the Paris region reveals the true position of the new towns in terms of office space actually constructed. As Table 9.1 indicates the annual level of planning

---

[(1)] See Dumbleton and Shutt (1979), for a strong presentation of the case against pension funds and their current activities and the position of the trades unions involved.

Table 9.1: Office Consents, Completions and Occupancy in New Towns of Paris Region 1975-1982 (square metres)

Consents for speculative office construction	89,214	20,700	57,769	21,608	120,170	165,608	138,550	249,311
% of Paris Region	30.7	16.0	49.1	22.1	28.7	34.4	36.0	55.9
Office Completions - speculative new and renovated	87,000	18,000	15,000	45,000	10,000	17,000	42,000	33,000
% of Paris Region	11.9	5.3	7.2	19.1	8.0	8.4	16.4	20.1
Occupancy of all new and renovated space	n/a	20,000	27,000	35,000	30,000	15,000	54,000	77,000
% of Paris Region		4.7	6.7	8.9	9.6	5.0	17.5	20.3

Source: (IAURIF 1983).

consents for speculative office building in the new towns has, with one exception, been above 20% of the total floor space granted consent in the period 1975-1982. By 1982, more than half of the speculative office floor space given planning consent annually was in the new towns. Yet despite the strong policies favouring the new towns, discussed in Chapter Six, the amount of offices actually constructed has been well below that for which permission has been granted. It is evident that many *agréments* (consents) have not been taken up in the new towns. Similarly if we examine the occupancy rate of all offices-both speculative and owner-occupied, we find that only by 1982 did the level of the new town's share rise above 20% of the Paris region's total. In contrast the department of Hauts de Seine, which includes La Défense had in 1981 and 1982, 39.1% and 35.2% respectively of speculative planning consents by area, whereas in the same years the department had 52.6% and 47.0% respectively of the completed space. (IAURIF, 1983).

The implication of these findings for policy-makers are of course very important. Only the very strongest control can actually effect the locational pattern required. The granting of planning permissions on a proportional locational basis is no guarantee of an eventual desired distribution. Certainly developers are much more likely to act on planning permissions gained in areas in high demand than secondary areas to which planners and others are endeavouring to direct development. But even then as discussed earlier, time lags in taking up permissions can distort any desired pattern.

A further option in terms of locational control is to institute a system of occupation permits either as an alternative or in addition to construction permits. Once again, the French have adopted this system, although it is not easy to assess its success since it is part of a system which as we have seen, includes other forms of control. In theory, however, a strong control of the occupation of office buildings according to demonstrable need would force occupiers to re-appraise their need for the most popular locations. It must be doubted, however, whether such a system could be instrumental in directing office occupiers to more peripheral locations.

The same could be said of a further policy alternative, that of a development tax. On a once-and-for-all basis, development taxes lack any real controlling mechanism. Such taxes are merely seen as a part of the development cost of an office property, reflected in the rent levels. More amenable to locational policy achievement are annual congestion taxes which shift the burden to the office occupier. Experience in central London, however, suggests that the locational attraction of such sites are so great that such extra costs would be absorbed. Certainly, as we saw in Chapter Five, rent differentials are already very considerable between the City of London and other locations, even in the relatively close proximity, yet office occupiers are willing to pay such premiums to stay in the City. Congestion taxes would need to be set at a very high level to have any appreciable impact. The major problem of political acceptability of such a measure would also remain to be solved.

POLITICAL CONSIDERATIONS

Whatever form of control is proposed, it cannot be effective without political will. Until the present socialist administration in France, there was some doubt concerning the real political commitment to a deconcentration of service activity from Paris. Certainly it was not to be at the expense of the city's role as a world financial capital. Indeed an aim of Giscard d'Estaing when Finance Minister in the late 1960s, was to create a financial capital to rival London. Similarly if the power of the financial institutions to invest wherever they choose is to be curbed, it will require a very much stronger political commitment than has been evident in Britain in the past. The investing vehicles i.e. the financial institutions, have strong defenders who are not always exclusively on the right wing of the political spectrum. Plender suggests that, "many conservatives believe that the best way to sell capitalism to the workers is through the medium of collective savings, whereby small sums are parcelled together by middlemen for investment in the stock market in larger, more economic blocks." (Plender 1982, p.30). Dumbleton and Shutt, more scathingly refer to the "thousands of small savers" justification for permitting activities of pension funds which they do not see as being socially desirable (Dumbleton and Shutt, 1979, p.336). On the subject of Labour party involvement in the rise of the financial institutions, Plender points to the irony of public corporations which Labour governments have created, spawning "some of the most powerful engines of modern finance capitalism, in the shape of nationalised industry pension funds - pension funds, moreover, which are legally bound to pursue objectives entirely independent of those of the public corporations whose employees they exist to serve." (Plender, 1982, p.30).

It is then difficult to see a British government acting strongly against the major institutions and curbing or directing their investment activities. Elsewhere, the major political considerations which have given rise to a curb on property investment activity has been those which led Canada and Australia to limit foreign ownership of property, although as pointed out in Chapter Three, their motivation stemmed from a more general desire to maintain domestic ownership of resources.

In the context of political considerations, it seems pertinent to re-examine Ambrose and Colenutt's prescription for resolving the problems of the redevelopment system (Ambrose and Colenutt 1975, pp.163-187). In rejecting any modification of existing legislation, they call for the restructuring of finance capital and the dissolution of the property market itself. They point to the inefficiencies of the operation of property market which amongst other excesses creates development in locations which are already congested. Daniels in the context of office development sees the practice of lending institutions favouring only established locations as one of "development inertia (leading) to locational inertia". (Daniels 1978, p.16). Certainly the process of mutually supportive decisions discussed in Chapter Two is a strong determinant of locational congestion. Ambrose and Colenutt state their objective to be, "to replace a property and planning system which respond to supply and demand with a development process which is organised around social needs and priorities." (Ambrose and Colenutt, 1975, p.164). To achieve this a number of specific proposals are made, some of which have already been subjected to initial consideration above.

Their most radical proposals may be summarised as: the establishment of a National Building Plan and of Public Development Corporations, the public control of finance capital and public ownership of all land and property, the restructuring of commercial rent to include rent subsidies or fair rents for commercial properties. They go on to suggest that the power of the property professions should be reduced and that planning should be based on progressively redistributive principles so aligning it with social policies aimed at reducing societal inequalities. The political nature of the planning process is also recognised in a proposal to make political appointments to Civil Service and local government. Amongst the other proposals are a call for greater public involvement and local community action and a better use of the educational system to promote a knowledge concerning urban issues in general. How far are these proposals feasible within our society? Certainly some have already been carried out, if only on a limited scale. Community involvement, for instance, was shown to be powerful in the Coin Street development controversy. Others present more difficulty and indeed could be said to be unrealistically idealistic without a more fundamental societal and political change.

Public Development Corporations have been established, such as the London Dockland Development Corporation, but it is equally true that such public bodies have rarely been able to approach urban development without a close scrutiny of economic viability. Perhaps more controversial is the notion of public ownership of finance capital. The French government has recently introduced the notion of compulsory loans from financial institutions, but beyond that the financial institutions have remained inviolate, despite the nationalisation of the banking sector. It is not difficult to see why this is the case generally when one considers the vested political interests in the financial sector discussed above. A large proportion of the population through either pension contributions or insurance policies wish to see maximum returns on capital invested. In Britain at least, that is sufficient to ensure freedom from political interference. Yet it is difficult not to agree with the view that such institutions are so powerful that they should now be subject to a degree of community accountability even if public ownership is unacceptable in political terms.

Public ownership of all land and property, thus effectively negating the land market may be the ideal solution, but in Britain even a degree of community involvement such as that proposed in the Community Land Act did not actually reach the statute books.

Intervention in the rent structure through a degree of rent control could serve as a redistributive measure in terms of the location of new office development. Certainly it would ensure that the financial institutions look beyond the narrow confines of the prime markets. At the same time, it must be acknowledged that as long as the British property industry is unconstrained with regard to its overseas activities, any such activity would merely stimulate a further withdrawal of finance capital to a more profitable and probably uncontrolled market overseas. The lack of concern by the property profession for social goals is unfortunate and is confirmed by even the most casual reading of professional property journals.
Ambrose and Colenutt go further to suggest that the professional bodies within the industry have a vested interest in maintaining both it and its property values, since their returns for professional services such as valuation work, are related to property values. Perhaps of equal concern is the indirect power which the professionals wield in terms of their advice to investors. They would be failing in their professional obligations if they were not to put the very highest priority on the economic returns to be gained from a particular property transaction or development. Many individuals, on the other hand, on whose behalf money is being invested, may prefer to see a decision taken with equal consideration being given to social goals and effects. There is of course no opportunity to gauge this particular preference.

In summary, Ambrose and Colenutt's proposals, some of which have now been partially met, are probably unrealistic, particularly in a western society which with only a few exceptions is more right-wing in political outlook than it was ten years ago. Nonetheless they highlight very effectively many of the problems which lie at the heart of the office development process and its relationship to the community at large.

FUTURE POLICY DIRECTIONS

Arising from the discussions above, it seems reasonable to call for a greater degree of public control over the office development process. Aside from the political difficulties, however, there are certain practical obstacles which make such control problematical. One such problem is that of determining market demand. Many would argue that the level of office development is determined by the demand from office users, although this book has frequently suggested that supply factors may be dominant over demand. The problem remains however, of determining demand and our past experiences suggest that this is no easy task. Returning to Schwartz's analysis of New York, he asserts that "a policy of government intervention assumes that government will somehow be better at predicting demand for office space than is the private sector... The efforts which have been made to date to predict both total demand and distribution of new construction have universally failed to hit the mark." (Schwartz 1979, p.232). With this activity as with any other land use, demand is subject to external pressures stemming from economic cyclical trends which may not be precisely predicted. Even a two year error can leave a large amount of office space either untenanted or unbuilt. The uncertainty surrounding the precise space requirements of offices adopting a new technology compounds an already difficult problem.

An aid to more accurate prediction is the careful monitoring of the operation of the process of development in all its aspects. In Britain at least, the data base amassed by public authorities is too weak to analyse the process in anything other than global and generalised terms. Ownership, source of finance, and type of occupancy of office space are all attributes which would assist in an accurate and valuable monitoring of the operation of the market. To this list, we may add rental levels since they may be seen as a reflection of the balance between supply and demand of office space. Much of this information can be gained in a piecemeal fashion, but only rarely can a comprehensive picture be built up. Intervention at any point in the development process is bound to be something of a lottery in the absence of detailed and reliable data bases.

IMPLICATIONS FOR GEOGRAPHICAL RESEARCH

In reviewing the field of office location research, Daniels comments that, "the influence of development companies on the location of office space is inextricably linked with the sources of finance upon which most speculative office development depends", and that "office location patterns are not simply a product of easily accessible opportunities for information gathering and exchange (communications) but are also determined by complex financial and other vested interests which are on both the demand and supply side, directly or indirectly, of the office market." (Daniels 1979, p.15). Research in the field of office location has been largely concerned with the role of communications and telecommunications in office organisation, the potential for and effects of relocation, and the regional development effects of office employment change. Yet Daniels' comment above is of crucial importance to a fuller understanding of office location. If the prime consideration for location is the economic welfare of the investor and developer rather than the end-user, then an analysis of office occupiers as a preliminary to a broad theory of office location is at best one-sided and at worst misleading.

At the macro-scale, a more general geographical understanding of the geography of finance capital is needed, particularly since such capital has now become so geographically dispersed. Such an area of research would cast further light not only on the patterns of property development, but also on economic development more generally. With only a few notable exceptions, however, geographers have preferred to side-step an analysis of the dynamics of this particular geographical distribution. Moving to the meso-scale, an understanding of the urban development process must be founded on an understanding of the property development system. Some geographers have already stressed this (see for instance Goodall 1972), but many basic texts in urban geography fail to recognise its importance, preferring to base explanations of urban land use and land use change on concepts of urban land economics which are far removed from the reality of the urban development process as reflected by the office development industry.

At the micro-scale, there is considerable scope for a geographical analysis of the decision-making processes leading to office development. The influence of individual property managers and developers is very considerable and their decisions have major consequences in transforming areas of cities. A geographical analysis of these decision-making processes, involving studies of perceptions of comparative investment opportunities would add considerably to our inadequate knowledge of the geography of the office development system.

BIBLIOGRAPHY

Alexander, I. (1979) *Office Location and Public Policy*, Longman, London

Ambrose, P. and Colenutt, B. (1975) *The Property Machine*, Harmondsworth: Penguin Books

APUR. (1980) *Schéma Directeur d'Aménagement et d'Urbanisme de la Ville de Paris*, Paris Projet, No.19-20, L'Atelier Parisien d'Urbanisme

Barras, R. (1979) *The Development Cycle in the City of London*, Centre for Environmental Studies Research Series 36

Barras, R. (1984) The Office development cycle in London, *Land Development Studies*, 1, 35-50

Bateman, M. (1971) *Some Aspects of Change in the Central Areas of Towns in West Yorkshire since 1945*. Department of Geography, Occasional Papers No.1, Portsmouth Polytechnic

Bateman, M. and Burtenshaw, D. (1979) The social effects of office decentralisation, in Daniels, P. (ed) *Spatial Patterns of Office Growth and Location*, Wiley, Chichester

BOURDAIS (1984) *Le Marché immobilier en France, Bureaux, Locaux d'activités et commerciaux*, Paris

Burtenshaw, D., Bateman, M. and Ashworth, G. J. (1981) *The City in West Europe*, Wiley, Chichester

CALUS (1983) *Property and Information Technology - the future of the offices market*, College of Estate Management, Reading

Cameron, G. C. (Undated) *Constraining the growth in employment of London, Paris and the Randstad - a Study of Methods*, (mimeograph)

Catalano, A. and Barras, R. (1980) *Office Development in Central Manchester*, CES Research Series 37, London

Colenutt, B. and Hamnett, C. (1982) *Urban Land Use and the Property Development Industry*, Unit 9, Urban Change and Conflict, The Open University, Milton Keynes

Comhaire, J. (1975) *L'Agglomeration de Bruxelles*, Notes et Etudes Documentaires Nos.4156-4157, Documentation Française

Conzen, M. R. G. (1960) *Alnwick, Northumberland: A study in Town Plan Analysis*, IBG Monograph No.27, London

Cowan, P. (1969) *The Office: A facet of urban growth*, Heinemann Educational, London

Cowan, R. (1982) Money to burn, *Town and Country Planning*, Vol.51, No.7, pp. 191-194

Crisp, J. (1982) Office Technology, Report in *Financial Times*, 19 March 1982

Damesick, P. (1979) Offices and inner-city regeneration, *Area*, 11, 41-47

Daniels, P. W. (1975) *Office Location: An Urban and Regional Study*, Bell and Sons Ltd., London

Daniels, P. W. (1979) Perspectives on Office Location Research, Chapter 1 in *Spatial Patterns of Office Growth and Location*, Ed. P. W. Daniels, Wiley, Chichester

Daniels, P. W. (1983) Modern technology in provincial offices: Some empirical evidence, *The Service Industries Journal*, 3(1), pp. 21-41

de Wandeleer, G. (1977) The Belgian Property Market, *Estates Gazette*, 242, 1977. pp. 32-33

Dicken, P. and Lloyd, P. (1981) *Modern Western Society*, Harper and Row, London

Digby, M. (1982) Office development and the micro-chip, *Chartered Surveyor*, March. pp. 445-446

Dumbleton, B. and Shutt, J. (1979) Pensions: The Capitalist Trap, *New Statesman*, Vol.98, 1979. pp. 334-337

Giuliano, V. E. (1982) The mechanisation of office work, *Scientific American*, 247(3), pp. 125-134

Goddard, J. B. (1975) *Office Location in Urban and Regional Development*, OUP, London

Goodall, B. (1972) *The Economics of Urban Areas*, Pergamon, Oxford

Hall, P. (1981) The geography of the fifth Kondratieff cycle, *New Society*, March 26th, pp. 535-537

Hall, R. (1972) The movement of offices from central London, *Regional Studies*, 6(4), pp. 385-392

Hampshire County Council (1972) *South Hampshire Draft Structure Plan*, Winchester

Hampshire County Council (1982) *Into the 1990s*, Winchester

Hennebury, J. (1982) The impact of microtechnology on landed property, *Estates Gazette*, Vol.262, 19th June, pp. 1151-1154

Hillier Parker (1983) *International Property Bulletin*, London

HMSO (1964) *South East Study; 1961-1981*, London

HMSO (1971) *Strategic Plan for the South-east*, London

HMSO (1982) *Business Monitor, Quarterly Statistics*, Business Statistics Office, London

HMSO (1983) *M4 Business Monitor. 1981 Overseas Transactions*, Business Statistics Office, London

Horwood, E. M. and Boyce, R. R. (1959) *Studies of the Central Business District and Urban Freeway Development*, University of Washington Press, Seattle

House, J. W. (1978) *France: an Applied Geography*, Methuen, London

IAURIF (1982) *Le Marché des Bureaux en Région de l'Ile de France en 1981*, Institut d'Aménagement et d'Urbanisme de l'Ile de France

IAURIF (1983) *Le Marché des Bureaux en 1982*, Institut d'Aménagement et d'Urbanisme de l'Ile de France

James, J. N. C. (1983) *Opportunities for the Smaller Corporate Fund Investor*, Paper delivered to Financial Times Conference on International Property Markets, January 19/20, London

Jones Lang Wootton. (1980) *Houston: an analysis of the City's economy and commercial real estate investment opportunities*, London

Jones Lang Wootton (1981) *New York: an analysis of the City's economy and commercial real estate opportunities*, London

Jones Lang Wootton (1983) *The Decentralisation of Offices from Central London: A CLOR Special Survey*, JLW Research, London

Kondratieff, N. O. (1978) The long waves in economic life, *Lloyds Bank Review*, No.129, pp. 41-60, (first published in English in Review of Economic Statistics, 1935)

LePage, A. E. Ltd. (1983) *Canadian Real Estate - 1984 (Market Survey)*, Toronto

Lean, W. and Goodall, B. (1966) *Aspects of Land Economics*, The Estates Gazette Ltd., London

Lewis, J. P. (1965) *Building Cycles and Britain's Growth*, MacMillan, London

Mallinson, M. H. (1983) *Institutions and Property - Fools and their money are soon parted?* Paper delivered to Financial Times Conference on International Property Markets, January 19/20, London

Massey, D. and Catalano, A. (1978) *Capital and Land: Land Ownership by Capital in Great Britain*, Edward Arnold, London

Moor, N. (1979) The contribution and influence of office developers and their companies on the location and growth of office activities in *Spatial Patterns of Office Growth and Location*, Ed. P. W. Daniels, Wiley, Chichester

Moseley, M. J. (1980) Strategic planning and the Paris agglomeration in the 1960s and 1970s: the quest for balance and structure, *Geoforum* 11(3), pp. 179-223

Murphy, R. E. and Vance, J. E. Jnr. (1954) A comparative study of nine central business districts, *Economic Geography*, 30, pp. 301-336

Nader, G. A. (1976) *Cities of Canada, Vol.II: Profiles of Fifteen Metropolitan Centres*, MacMillan of Canada, Toronto

Noel Alexander Associates (1983) *New Foreign Bank Offices in London - 1982*, London

Parkes, D. N. and Thrift, N. J. (1980) *Times, Spaces and Places*, Wiley, Chichester

Plender, J. (1982) *That's the Way the Money Goes*, André Deutsch: London

Price, S. G. (1979) *Introducing the Electronic Office*, National Computing Centre: London

Pye, R. (1977) Office location and the cost of maintaining contact, *Environment and Planning A*, 9, pp. 149-168

Pye, R. (1979) Office location: the role of communications and technology in *Spatial Patterns of Office Growth and Location*, Ed. P. W. Daniels, pp. 239-276, Wiley, Chichester

Pye, R. and Williams, E. (1977) Teleconferencing: is video valuable or is audio adequate? *Telecommunications Policy* 1, pp. 230-241

Royal County of Berkshire (1977) *Structure Plan: Report of Survey*, Reading

Sant, M. (1973) *The Geography of Business Cycle*, LSE Geographical Papers, No.5, London

Scargill, D. I. (1983) *Urban France*, Croom Helm, London

Schwartz, G. G. (1979) The office pattern in New York City, 1960-1975, Chapter 9, in Daniels, P. W. *Spatial Patterns of Office Growth and Location*, Wiley

Short, J., Williams, E. and Christie, B. (1976) *The Social Psychology of Telecommunications*, Wiley, London

Stevenson, D. (1981) *Westminster's Limited Contribution to the 'New Office Boom'*, Paper presented to the first CES London Conference, The Office Boom in London, 9 February 1981

Strohm, P. (1983) Hi-tech challenge for offices, *Estates Gazette*, Vol.266, 16 April, p. 198

Taylor, M. and Thrift, N. (eds), (1982) *The Geography of Multi-Nationals*, Croom Helm, London

Tuppen, J. (1983) *The Economic Geography of France*, Croom Helm, London

Vickers da Costa (1983) *United Kingdom Research Report*, Property Sector Review 23, London

Whitehand, J. W. R. (1972), Building cycles and the spatial pattern of urban growth, *IBG Transactions*, No.56, pp. 39-56

Whitehand, J. W. R. (1983) Land use structure, built form and agents of change, Chapter 3 in *The Future for the City Centre*, IBG Special Publication No.14, Edited by R. L. Davies and A. G. Champion, Academic Press, London

Yannopoulos, G. (1973) Local income effects of office relocation, *Regional Studies*, 7, pp. 33-46

Yeates, M. (1975) *Main Street*, Macmillan, Toronto

INDEX

For Product Safety Concerns and Information please contact our EU representative GPSR@taylorandfrancis.com
Taylor & Francis Verlag GmbH, Kaufingerstraße 24, 80331 München, Germany

www.ingramcontent.com/pod-product-compliance
Lightning Source LLC
LaVergne TN
LVHW010628300926
849162LV00001B/94
* 9 7 8 1 0 3 2 0 0 5 3 8 6 *